IUTAM SYMPOSIUM ON LUBRICATED TRANSPORT OF VISCOUS MATERIALS

FLUID MECHANICS AND ITS APPLICATIONS
Volume 43

Series Editor: R. MOREAU
MADYLAM
Ecole Nationale Supérieure d'Hydraulique de Grenoble
Boîte Postale 95
38402 Saint Martin d'Hères Cedex, France

Aims and Scope of the Series

The purpose of this series is to focus on subjects in which fluid mechanics plays a fundamental role.

As well as the more traditional applications of aeronautics, hydraulics, heat and mass transfer etc., books will be published dealing with topics which are currently in a state of rapid development, such as turbulence, suspensions and multiphase fluids, super and hypersonic flows and numerical modelling techniques.

It is a widely held view that it is the interdisciplinary subjects that will receive intense scientific attention, bringing them to the forefront of technological advancement. Fluids have the ability to transport matter and its properties as well as transmit force, therefore fluid mechanics is a subject that is particulary open to cross fertilisation with other sciences and disciplines of engineering. The subject of fluid mechanics will be highly relevant in domains such as chemical, metallurgical, biological and ecological engineering. This series is particularly open to such new multidisciplinary domains.

The median level of presentation is the first year graduate student. Some texts are monographs defining the current state of a field; others are accessible to final year undergraduates; but essentially the emphasis is on readability and clarity.

For a list of related mechanics titles, see final pages.

IUTAM Symposium on

Lubricated Transport of Viscous Materials

Proceedings of the IUTAM Symposium
held in Tobago, West Indies,
7–10 January 1997

Edited by

HAROLD RAMKISSOON

Department of Mathematics and Computer Science,
The University of the West Indies,
St Augustine, Trinidad

SPRINGER-SCIENCE+BUSINESS MEDIA, B.V.

A C.I.P. Catalogue record for this book is available from the Library of Congress.

ISBN 978-94-010-6208-4 ISBN 978-94-011-5248-8 (eBook)
DOI 10.1007/978-94-011-5248-8

Printed on acid-free paper

Originally published by Kluwer Academic Publishers in 1998
Softcover reprint of the hardcover 1st edition 1998

TABLE OF CONTENTS

Suspensions

Miscellanous

PREFACE

The main objective of the First International Symposium on Lubricated Transport of Viscous Materials was to bring together scientists and engineers from academia and industry to discuss current research work and exchange ideas in this newly emerging field. It is an area of fluid dynamics devoted to laying bare the principles of the lubricated transport of viscous materials such as crude oil, concentrated oil/water emulsion, slurries and capsules. It encompasses several types of problem. Studies of migration of particulates away from walls, Segré-Silverberg effects, lubrication versus lift and shear-induced migration belong to one category. Some of the technological problems are the fluid dynamics of core flows emphasizing studies of stability, problems of start-up, lift-off and eccentric flow where gravity causes the core flow to stratify. Another category of problems deals with the fouling of pipe walls with oil, with undesirable increases in pressure gradients and even blocking. This study involves subjects like adhesion and dynamic contact angles. The topics of shear-induced diffusion of small particles and wall slip in slow flow are other appropriate subjects. Computer-intensive studies of flow-induced microstructures and moving interface problems are yet additional research directions.

The general consensus was that the Symposium was a tremendous success, although the number of presentations fell below expectations. Scientists from the petroleum industry, and this includes INTEVEP (Venezuela), Schlumberger and Syncrude Canada Ltd. , and consultants to oil companies actively participated in the Symposium. The meeting produced new insights which should lead to further interesting research work and established contacts for possible joint investigations.

This book, the Proceedings of the Symposium, represents some of the finest work to date in the field. The contributors are among the most innovative and imaginative. Most unfortunately, however, is the fact that some of the papers presented were, for one reason or another, not submitted for publication in the Proceedings. This is a disappointment to the organizers of the meeting and, I am sure, to those researchers who were unable to attend the meeting but nonetheless are looking forward to the release of the Proceedings.

I wish to thank all those who made this First Symposium a success, and this includes my co-Chairman Prof. D. Joseph, the contributors, the sponsors (particularly the University of the West Indies who made this publication possible), the members of the Scientific Committee and the members of the Organizing Committee - Dr. T. Jagai, Dr. W. Mellowes, Dr. K. Rahaman and Mr. G. Ramdath. Special mention must be made of the warm hospitality extended to us by the Tobago House of Assembly for which we are grateful.

August 6, 1997
Harold Ramkissoon.

ACKNOWLEDGEMENTS

The Editor gratefully acknowledges the financial support of sponsors, local companies and institutions that made possible this "International Symposium on Lubricated Transport of Viscous Materials" held in Tobago from January 7-10, 1997 that resulted in the publication of this Proceedings. In particular we thank the following :

a) International Union of Theoretical and Applied Mechanics

b) The University of the West Indies

c) The Caribbean Congress of Fluid Dynamics and

d) The Tobago House of Assembly.

INTRODUCTION

Lubricated transport of viscous materials is today becoming one of the more applicable areas of Fluid Dynamics. Although as a field with underlying principles it is relatively new, the concept of lubrication with its associated theory has been around some time. And, of course, one of the early applications has been in journal bearing. The interest in this new field is due mainly to its application in the petroleum industry. However, it is certainly not restricted to the transport of oil. Research work, for example, is being undertaken in the transportation of coal logs.

This Proceedings is a good reflection of the current research work being done on "Lubricated Transport of Viscous Materials". In the first section of the book D. Joseph, a leading expert in the field, gives an introduction , punctuated with illustrations. The topics include "Types of Lubricated Flows" and "Melt Fracture and Snakeskins".

The second section covers flows through pipes. There are some notable contributions here in the areas of core-annular flow and transport of capsules. Parafin deposition in multiphase flowlines and well bores is a major problem encountered in the petroleum industry. A Joint Industrial Project was initiated in 1995 at the University of Tulsa to investigate this problem. The first part of this project is presented here.

In the third section, the focus is on suspensions and the contributions here are interesting. The work titled "Thoughts Old and New on Moving Contact Lines" in the final section is very thought provoking.

I wish to express my deep appreciation to all those who made the publication of this Proceedings possible.

August 6th, 1997,
Dr. H. Ramkissoon.

LUBRICATED TRANSPORT OF VISCOUS MATERIALS

Lecture presented at IUTAM conference on lubricated transport of viscous materials, Tobago, Jan. 7, 1997.

DANIEL D. JOSEPH
University of Minnesota
Department of Aerospace Engineering & Mechanics
107 Akerman Hall, 110 Union Street
Minneapolis, MN 55455

Contents

1. Phase arrangements

In two phase flows the dynamic response is tied to the way the phases are arranged. Many configurations are realized in practice; these are often described by flow charts for

- Liquid-liquid
- Gas-Liquid
- Liquid-Solid

In figure 1 you see a cartoon of flow types in oil-water flows. A flow chart for gas-liquid flows is shown in figure 2. There are definitely flow types in liquid-solid flows (figure 3), but as far as I know, flow charts have not yet been prepared for liquid-solid flows.

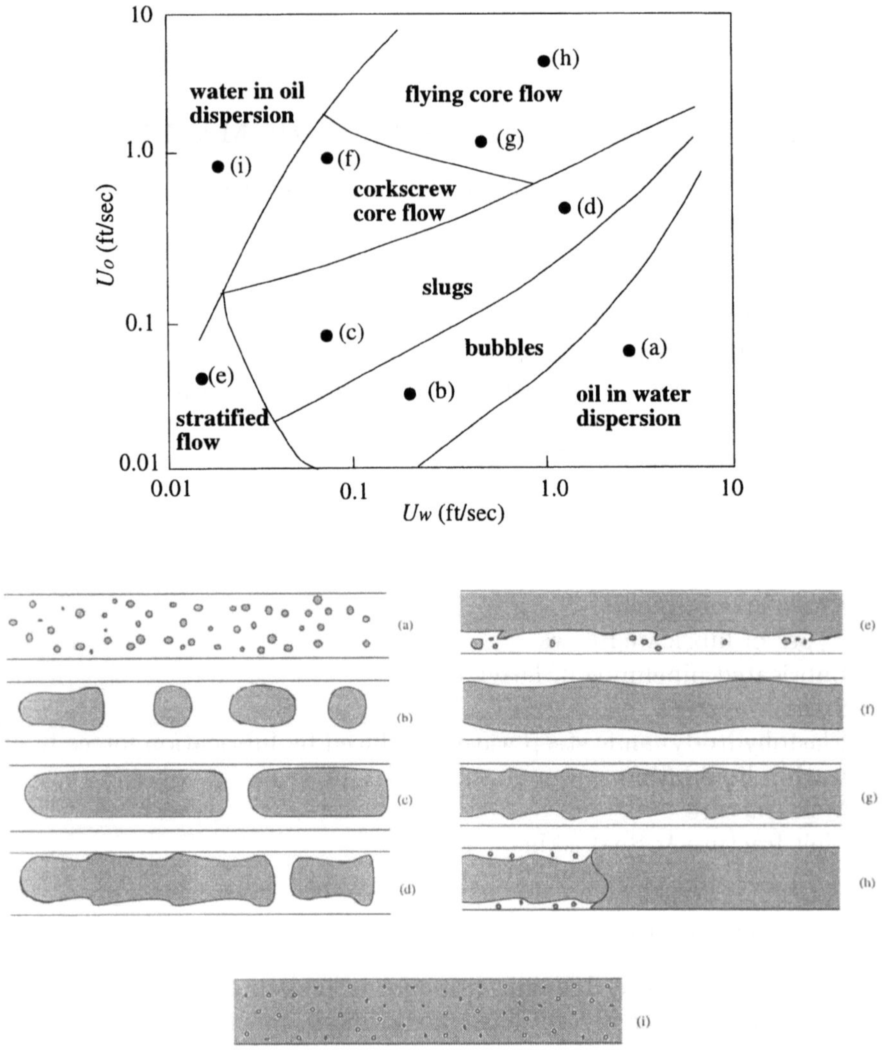

Figure 1. (Joseph, *et al.*, 1997) Cartoons of a flow chart and flow types in horizontal flow when the oil is lighter. The flow is from right to left. All flows but (e) and (i) are lubricated. The holdup ratio $h = \overline{U}_0/\overline{U}_w$ where $\overline{U}_0$ is the average oil velocity and $\overline{U}_w$ is the average water velocity. $h = 1$ for finely dispersed flow like (i), $h = 2$ for perfect core flow, $h \approx 1.5$ for wavy flow.

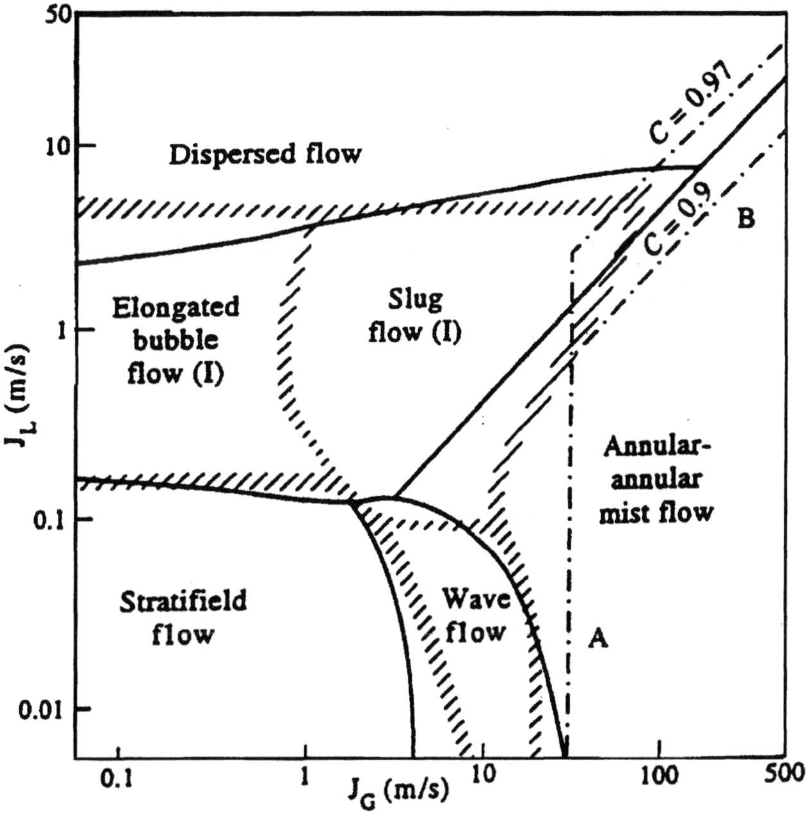

Figure 2. (Joseph, Bannwart and Liu, 1996) Taitel and Dukler's (1976) flow map of horizontal air-water flow at 1 bar and 25°C. The pipe diameter is 25 mm. The solid lines stand for the theoretical transitions proposed by Taitel and Dukler. The dashed lines indicated the experimental diagram of Mandhane, Gregory and Aziz (1974). $\mu_{eff,G} = \mu_{eff,L}$ on the dashed-dot line.

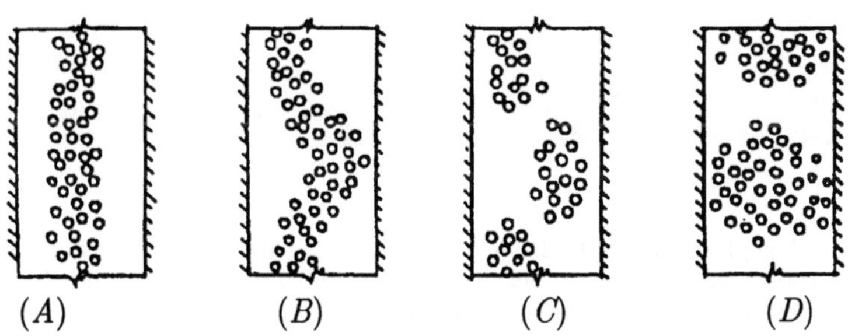

Figure 3. (Brandt and Bugliarello, 1996) Two-dimensional shear flow of neutrally bouyant rigid spheres. Development of instabilities. (*A*) Concentrated Core; (*B*) waves; (*C*) separation; (*D*) discontinuous flow.

2. Types of lubricated flows

- Solids lubricated by liquids.
    - journal bearing
    - pneumatic tube
    - coal log (Henry Lui)
- Lubricate a very viscous liquid with a mobile liquid, say water
    - core flow
    - rollers (figure 4, Joseph *et al.*, 1986)

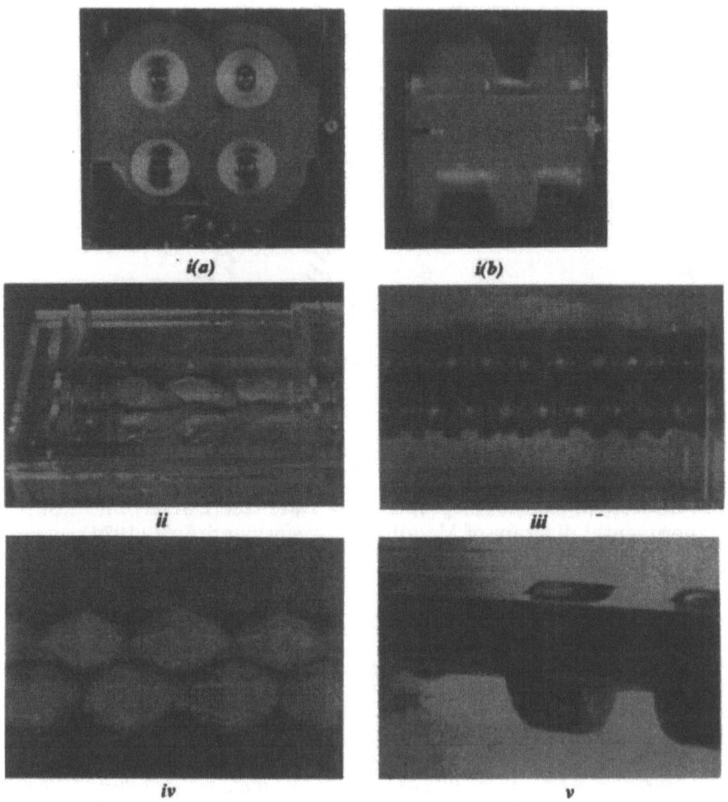

Figure 4. (Joseph *et al.*, 1986) Rollers. STP rotating in water, lubricated all around. The rollers rotate as a rigid body because they are lubricated. You would think they are solids, but they collapse when the motor is turned off. It is amusing to think that we can create an effective solid from a fluid by lubrication.

– Lubricate slurries with water (the particles can't be too heavy or the flow too slow). The water migrates to the wall and the solids to the core (figure 3). Lubricated heavy oil in water emulsions (Orimulsions, Nez *et al.*, 1996).
– Lubricate turbulent gas (high eddy viscosity) with water. Annular gas-liquid flow in horizontal pipes (figure 5) and up-flow.

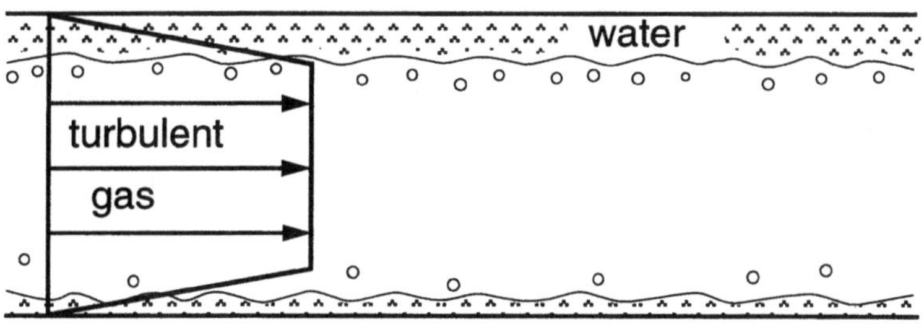

Figure 5. Horizontal annular gas-liquid flow gets stable when the eddy viscosity in the gas gets to be as large as the water (· — · — · line of Joseph, Bannwart and Liu (1996) in figure 2).

– Lubricate turbulent gas in up-flow with a concentrated dispersion of solids raining down pipe walls (figure 6). The gas core is again stabilized by an increase in the eddy viscosity of gas beyond the effective viscosity of the compacted solids. In lubricated slurries particles go to the core, just the opposite.

3. Lubricated pipelining of heavy oil

– Water will go to the pipe wall if the oil viscocity is larger than about 5 poise.
– Drag reductions of the order of the viscocity ratio are possible

$$\mu_0/\mu_w = 1000/\tfrac{1}{100} = 10^5$$

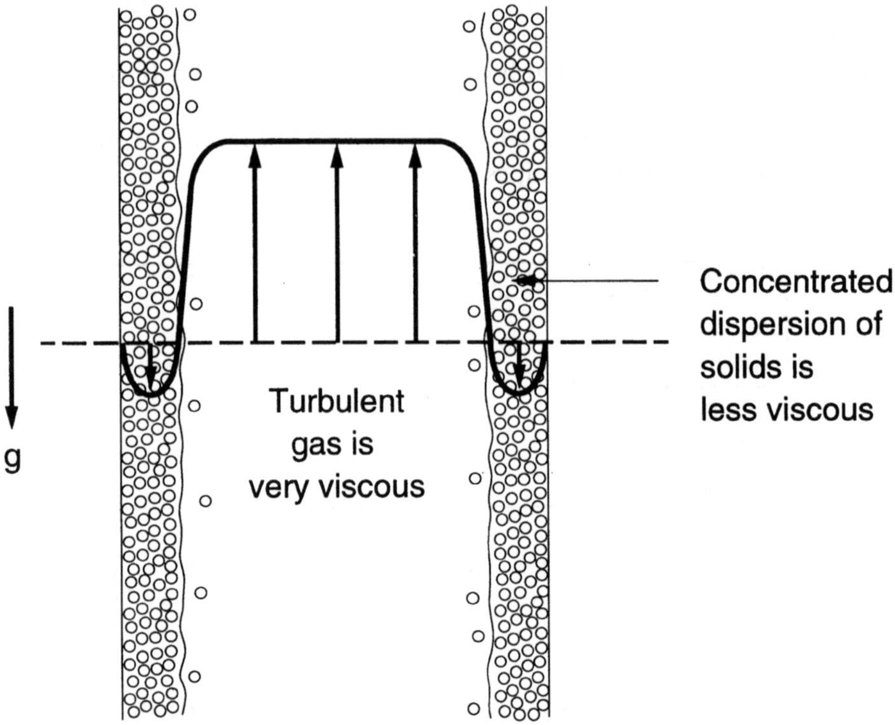

Figure 6. (Joseph, Bannwart and Lui, 1996) Annular solid-gas flow. It is strange to think concentrated solids on the wall have less viscosity than the turbulent gas and that the solids are a lubricant as is suggested by theory (Iske *et al.*, 1995; Joseph, Bannwart and Lui, 1996)

3.1. HISTORY (SEE JOSEPH AND RENARDY, 1993, JOSEPH *ET AL.*, 1996)

1906 Isaacs and Speed proposed to rifle the pipe and throw the water to the wall by centripetal acceleration
1930's Various oil companies built short pilot lines, a few miles
1960's Canadian studies at Alberta by Govier, Hodgson, Charles
1970 - 1982 Shell ran a commercial 8″ line from Bakersfield to the upgrading facility at ten section. Ran especially well when using produced water, but you got lubrication only for flow velocities

$$U > 3 \text{ ft/sec}$$

You need inertia to levitate the flow
1996 Syncrude froth pilot: This is a 1 km long 24″ pipe. You get self-lubrication rather than lubrication because the water is already there.

The water is a colloidal dispersion of clay which promotes the self lubrication. Additional water is not added.

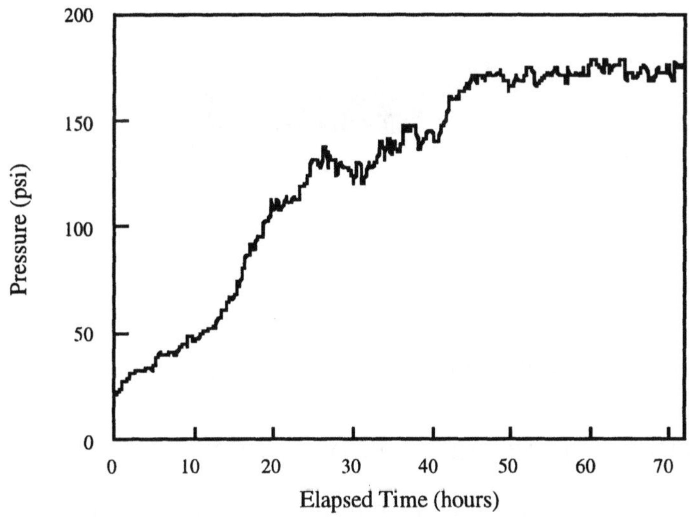

Figure 7. (Joseph *et al.*, 1996) Fouling of the San Tomé test loop with Zuata crude. input fraction = 4%, superficial oil velocity = 1.5 m/sec. Pressure losses increase monotonically as the pipeline fouls. High blockage was experienced after 2 1/2 days of operation.

3.2. FOULING

This is a serious problem for some oils and pipe walls. Venezuelan oils have a high concnetration of asphaltenes which are apt to stick to carbon steel pipes. This problem has so far frustrated the development of commercial lubrication in Venezuela (figure 7). Research in anti-fouling has been carried out with cement-lined and polymeric-lined pipes.

3.3. TURBULENCE

Figure 8 gives friction factor vs. Reynolds number (λ vs. $\mathbb{R}$ defined in Arney *et al.*, 1993) for all published core flow data. The oil moves nearly as a rigid body; the water is usually turbulent and follows the Blasius correlation $\lambda = 0.316/\mathbb{R}^{0.25}$ for $\mathbb{R} < 10^6$. The scatter on the high side of turbulent water in the gap is due to fouling. You can derive this more or less from a K-E model for pure water, no fitting, assuming a smooth interface (Huang *et al.*, 1994).

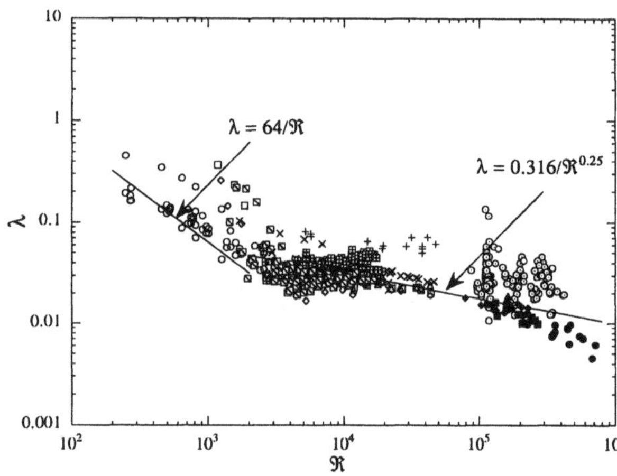

Figure 8. (Arney *et al.*, 1993) Friction factor vs. Reynolds number from different sources.

3.4. CREATION AND ANNIHILATION OF FOULING

The critical cause for concern is the build-up of fouling. In fact most oil will stick to the walls of a carbon steel pipe, but there is no buildup because eventually the oil is torn off the fouled pipe as fast as it is put on. We saw this kind of response in our studies of number 6 fuel oil. There was a layer of oil on the wall, but no build-up of fouling, the pressure drop did not increase.

The same balance between creation and annihilation of fouling occurs for bitumen froth (figure 9).

4. Steep Waves (Joseph and Lui, 1996, Joseph, 1997)

When fluid flows through a gap defined by a wave crest and a solid wall, the pressure at the wave front is much larger than at the rear. This leads to wave steepening at the front side of the wave. This kind of wave steepening appears to be ubiquitous, not only in lubricated transport but also in material processing.

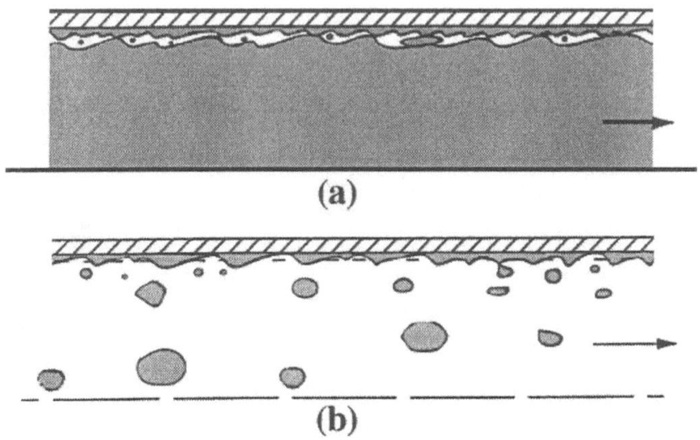

Figure 9. Creation and annihilation of fouling of bitumen froth. The froth is weak because it is covered by clay from clay water and can't stick to itself. The water can coelesce to form a lubricating sheath between the fouled wall and the core. Bits of oil are torn away and new fouling appear to be in balance, since the pressure gradient doesn't increase. You pay a price for fouling approximately 20 times the cost of water alone. You can wash the froth from the pipe with clay water (b); pegging is not required.

4.1. INSTABILITY OF LONGWAVES

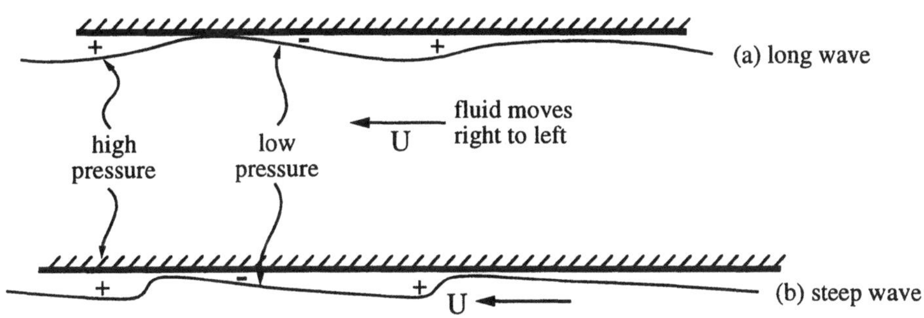

Figure 10. The wave front (a) steepens (b) due to high pressure at the front.

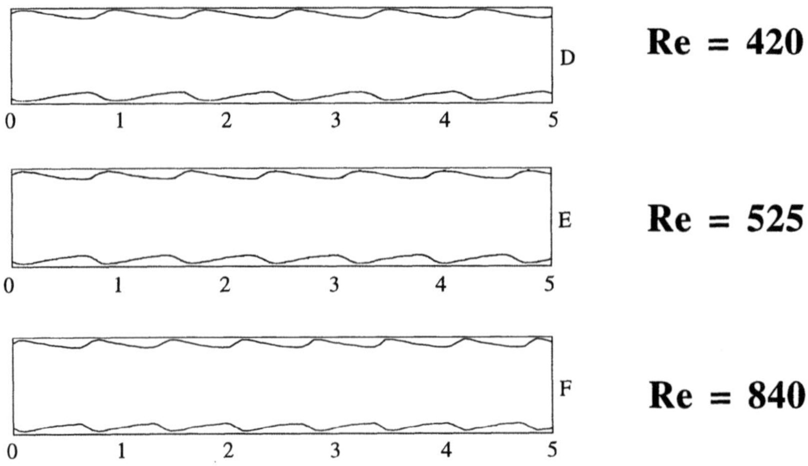

Figure 11. Numerical calculation of Bai *et al.*, (1996) for $[\eta, h] = [0.86, 1.4]$. The wave length shortens and the wave front steepens as the Reynolds number is increased.

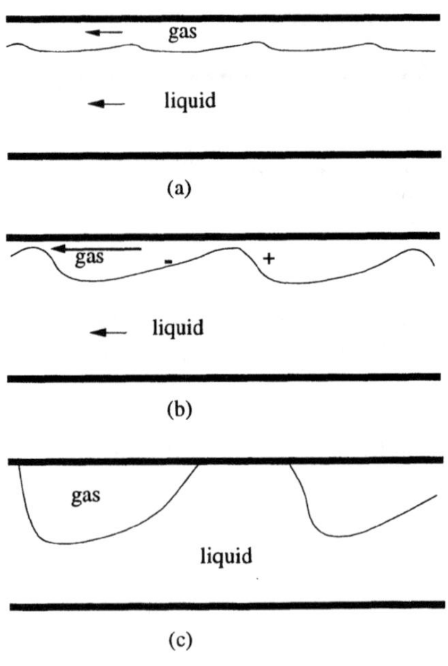

Figure 12. (Joseph *et al.*, 1986) Slugging in gas-liquid flow. The high pressure at the front side steepens the wave and the low pressure at the back side smoothes it. If the amplitude of the wave is large enough it will touch the wall. Bernoulli effects can also suck the liquid to the wall when the gas velocity is large. Note that we have avoided drawing the ripples and small roll waves which are probably always superimposed on the gas-liquid interface. We focus on macro-structures.

4.2. LEVITATION OF A SLIPPER BEARING

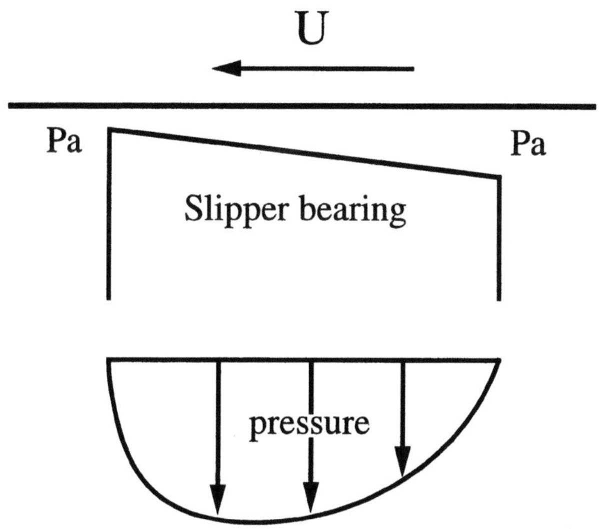

Figure 13. The pressure distribution levitates the bearing. If U is reversed, the slipper bearing will be sucked toward, rather than pushed away from the moving wall.

4.3. LEVITATION OF CORE FLOW

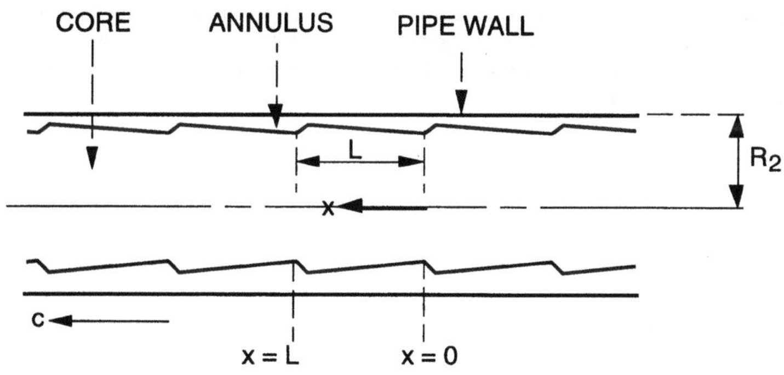

Figure 14. (after Ooms *et al.*, 1984) The core is at rest and the pipe wall moves to the left. The core shape is assumed to be a sequence of slipper-like bearings. This configuration should not levitate in the low Reynolds number of lubrication theory; you get anti-levitation instead, suggesting that inertia is needed for levitation.

4.4. STEEP WAVES IN ANNULAR FLOWS

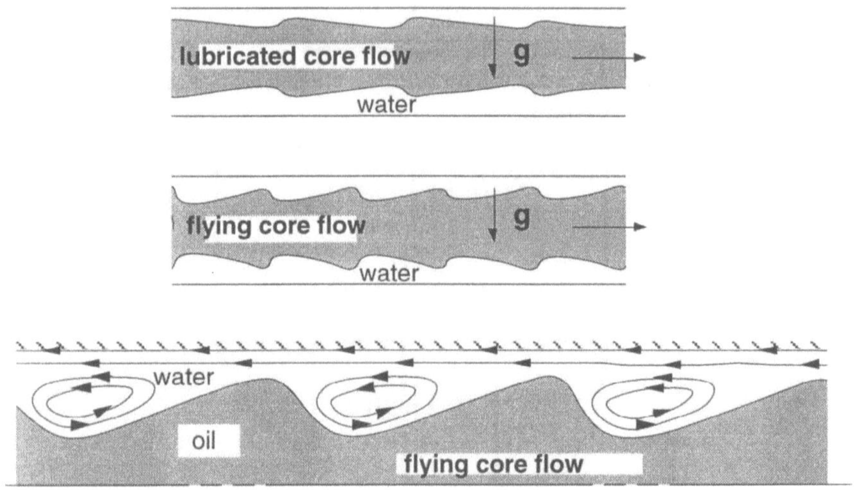

Figure 15. (Feng, *et al.* 1995) (*top*) The interface resembles a slipper bearing with the gentle slope propagating into the water; the shape of these waves is unstable. (*middle*) The high pressure at the front of the wave crest steepens the interface and the low pressure at the back makes the interface less steep. (*bottom*) The pressure distribution in the trough drives one eddy in each trough. The waves in (a) are unstable and lead to (b). To get a lift from this kind of wave it appears that we need inertia, as in flying. Liu's (1982) formula for capsule lift-off in a pipelin in which the critical lift off velocity is proportional to the square root of gravity times the density difference is an inertial criterion. Industrial experience also suggests an inertial criterion, since CAF in the Shell line could be maintained only when the velocity was greater than 3 ft/s; at lower velocities the drag was much greater.

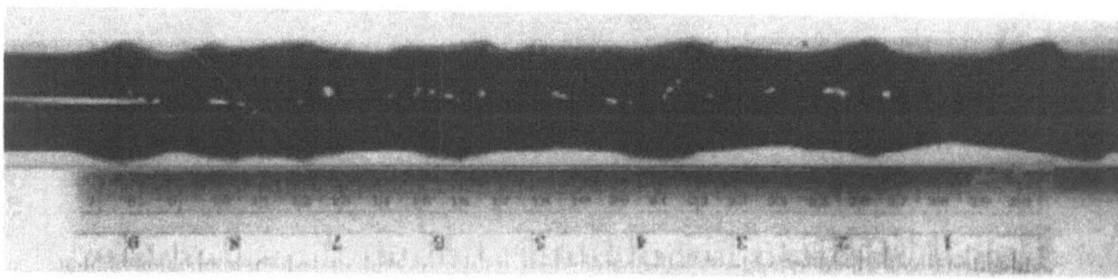

Figure 16. (Joseph, 1997). Core annular flow of #6 fuel. The saw-tooth waves on the oil core in a horizontal pipeline. The flow is from left to right.

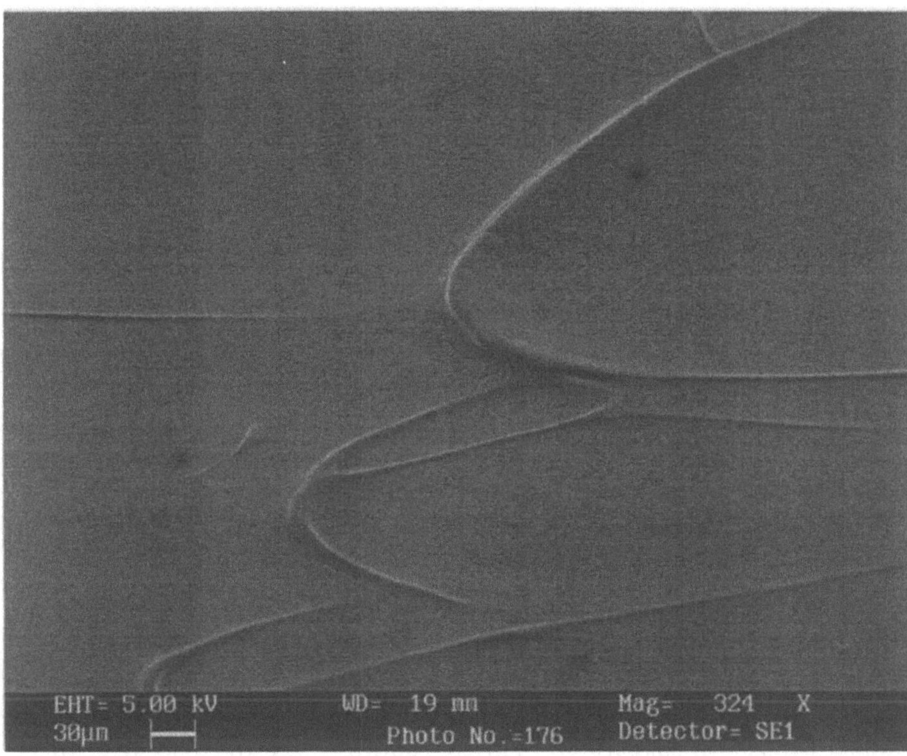

Figure 17. (Joseph, 1997, Courtesy of R.P.G. Rutgers) Film blown of a 40mm Kiefel extruder with a 0.8 mm die gap width and wall shear rate = 277 s^{-1}. Direction of melt flow is right to left. Steep fronts advance.

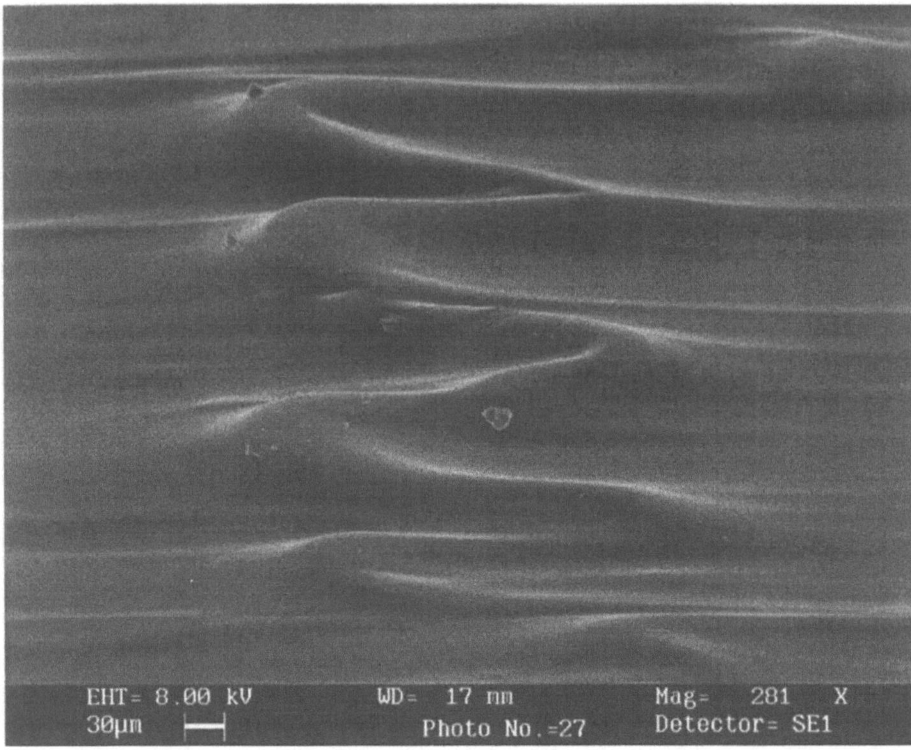

Figure 18. (Courtesy of R.P.G. Rutgers) Extruded tape from a 25 mm Betol single screw extruder with a 15:1 abrupt entry slit die of 1 mm gap width (x), and 8 mm length (z). Haul off ratio of 7.3:1, melt temperature of 178°C, wall shear rate = 163 s^{-1}. The extrudate surface shown is the y-z plane. Direction of flow from right to left. Scale as indicated on the photograph.

5. Elastohydrodynamic Steep Waves Produced by Lubrication Forces in Ultrathin Liquid Films

Figures 22, 23, 24 are photographs of the consequences of ultrathin film lubrication dynamics on the mica boundaries of a surface force apparatus by Israelichvili and co-workers (Kuhl *et al.*, 1994). The boundaries of the thin film are smooth mica. The fluid is polybutadiene $M \approx 10^4, \mu = 180$ poise. Typical values are velocity 10^{-4}cm/sec, gap $h = 2 \times 10^{-6}$cm, $\mathbb{R} \approx 10^{-12}$. This is Stokes flow, but it gives rise to steep waves of mica due to high and low pressures associated with lubrication.

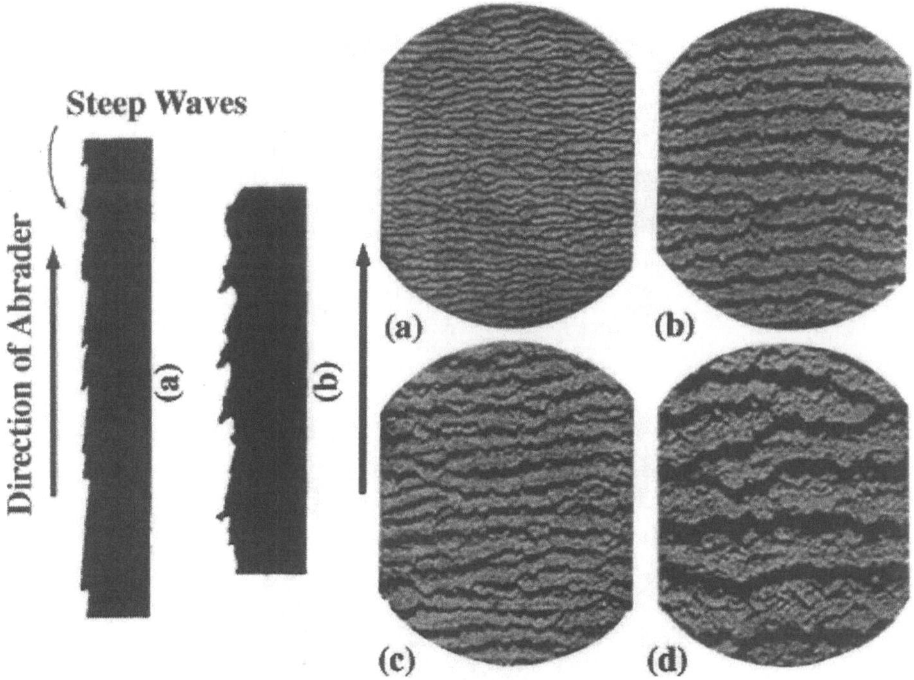

Figure 19. (left): Profiles of abrasion patterns: (a) gum vulcanizate of NR abraded under 1 · 6 kg/cm^2 on silicon-carbide cloth, 5×; (b) worn NR tyre surface, horizontal magn. 30×, vertical magn 42 · 5×. Trans IRI., 1952, 28, 259, Fig 3.

Figure 20. (right): Abrasion patterns on two black-filled NR vulcanates: (a) and (c) 45 pph HAF; (b) and (d) 25 pph HAF. Tracks: (a) and (b) fine tarmac; (c) and (d) coarse concrete. Direction of abrasion upwards. Magn. 14 ×. Wear, 1958, 1, 406, Fig. 17.

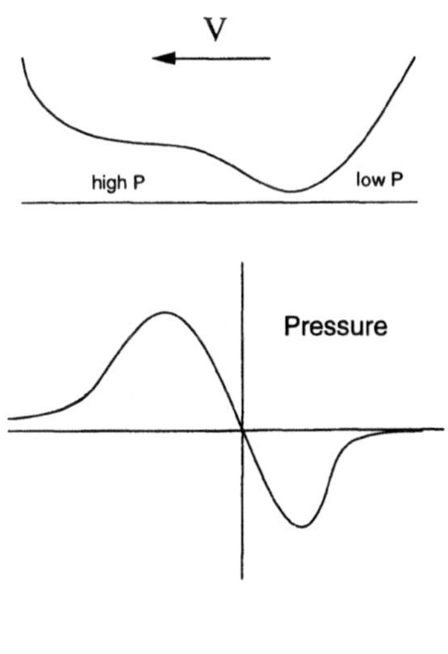

Figure 21. Pressure distribution according to lubrication theory

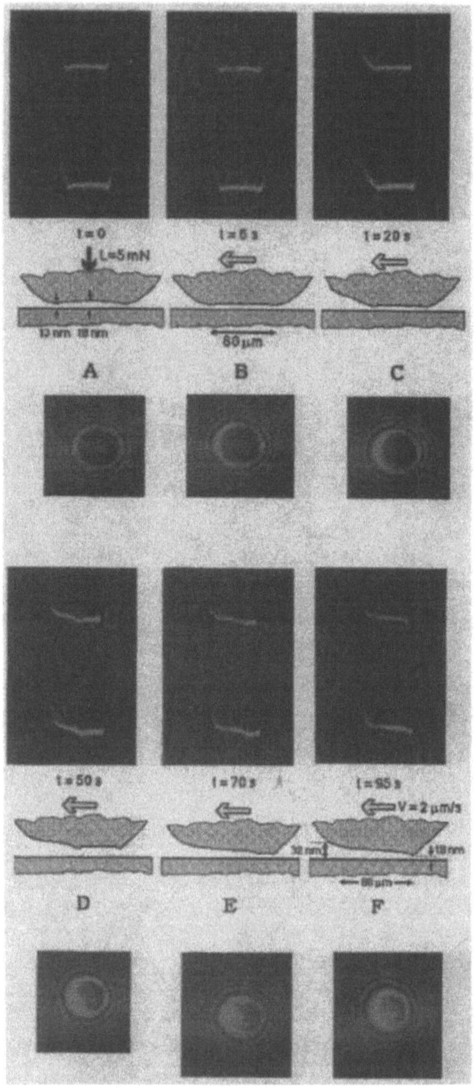

Figure 22. (Kuhl *et al.*, 1994) Development in time of the elastohydrodynamic deformation of shearing surfaces, starting from the profile at rest (A at $t = 0$) and gradually changing to the steady-state profile (F) at a constant sliding velocity of 2μm/sec. In the optical microscope view (bottom) the circular patterns (Newton's rings) arise from constructive interference of monochromatic light. The dark central region is the area of minimum gap separation. The gradully changing film thickness profile as the surfaces reach steady sliding is seen as a changing brightness within the contact zone.

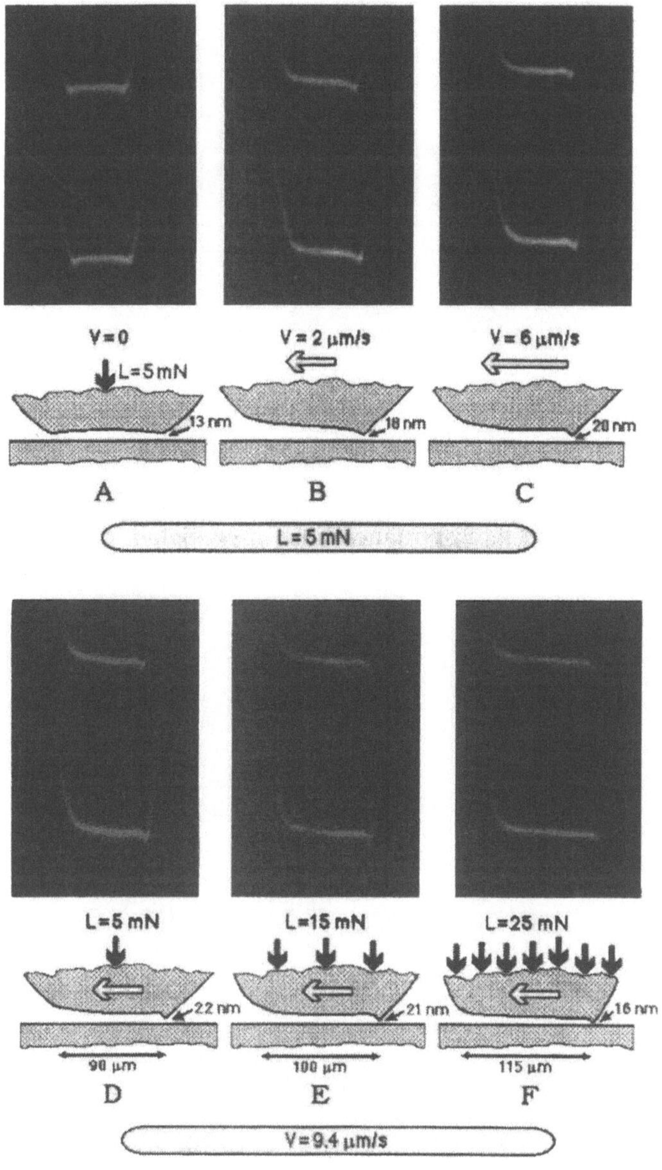

Figure 23. (Kuhl *et al.*, 1994) (A-D) The effect of increasing velocity on surface deformation at constant load L. The deformation of the front edge and the separation of the two surfaces increases with increasing velocity. (D-F) At constant sliding velocity, an increased load decreases the film thickness in the gap and increases the area of contact.

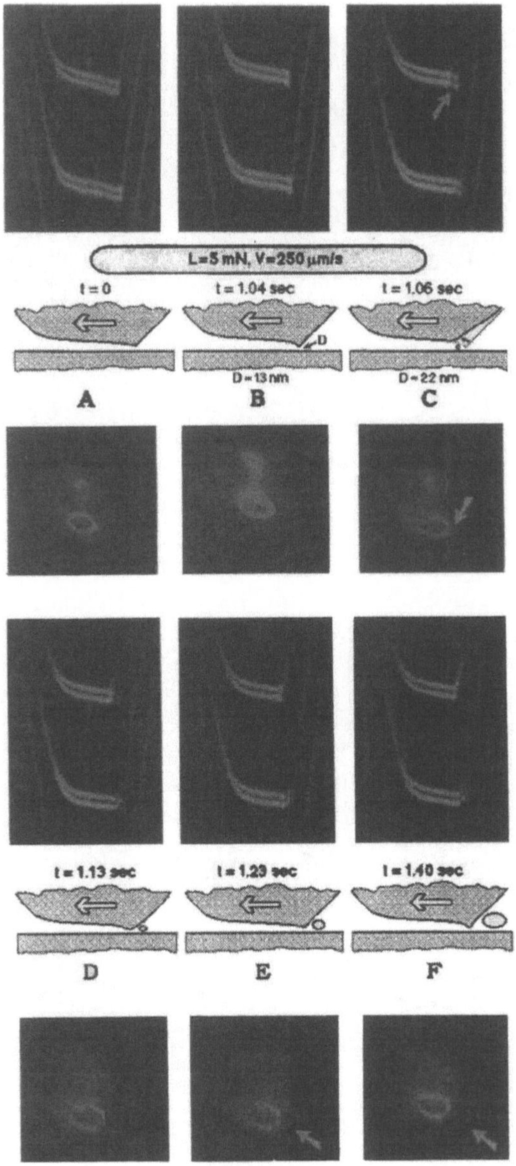

Figure 24. (Kuhl *et al.*, 1994) Surface deformation and formation of a cavity at high sliding velocity, $\eta > \eta_c$. The trailing edge (B) becomes very pointed due to the large tensile stresses developed under high shear rates. A cavity is formed as the surfaces rapidly snap forward (C) to a less deformed shape. Pictures D to F show the evolution and growth of a cavity with time.

6. Melt Fracture

These are cracks, fractures or waves on polymer extrudate, plastics. At low rates of extrusion there is no extrudate distortion; at a critical extrusion the polymer is said to slip. I think that a layer near the wall becomes soft, a lubrication layer (Joseph and Lui, 1996, Joseph, 1997). You see the onset of lubrication as a break in the flow curve.

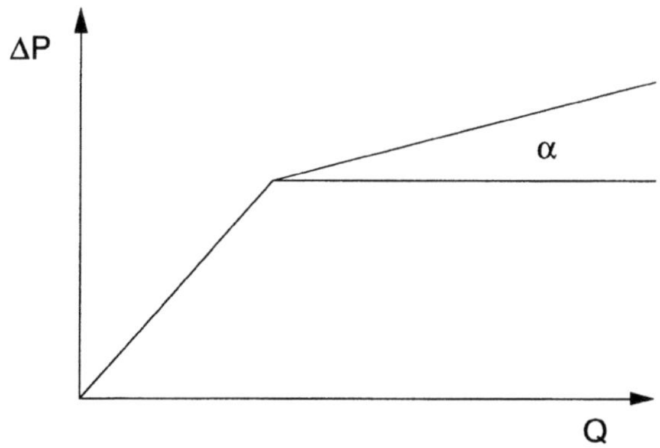

Figure 25. Cartoon of the flux curve given the pressure drop across a capillary versus the volume flux. The frictional resistance **decreases** at the point of discontinuity, with the complete slipping when $\alpha = 0$.

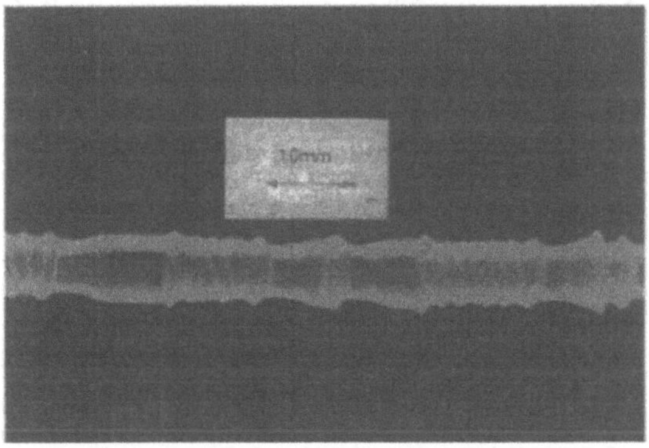

Figure 26. (Kalika and Denn, 1987) Extrusion of LLDPE: Transition region observed in stick-slip flow region $8V/D = 985, L/D = 100/1$. The authors do not remember which way the melt was extruded. We guess that the steep wave advances

7. Melt Fracture & Snakeskin

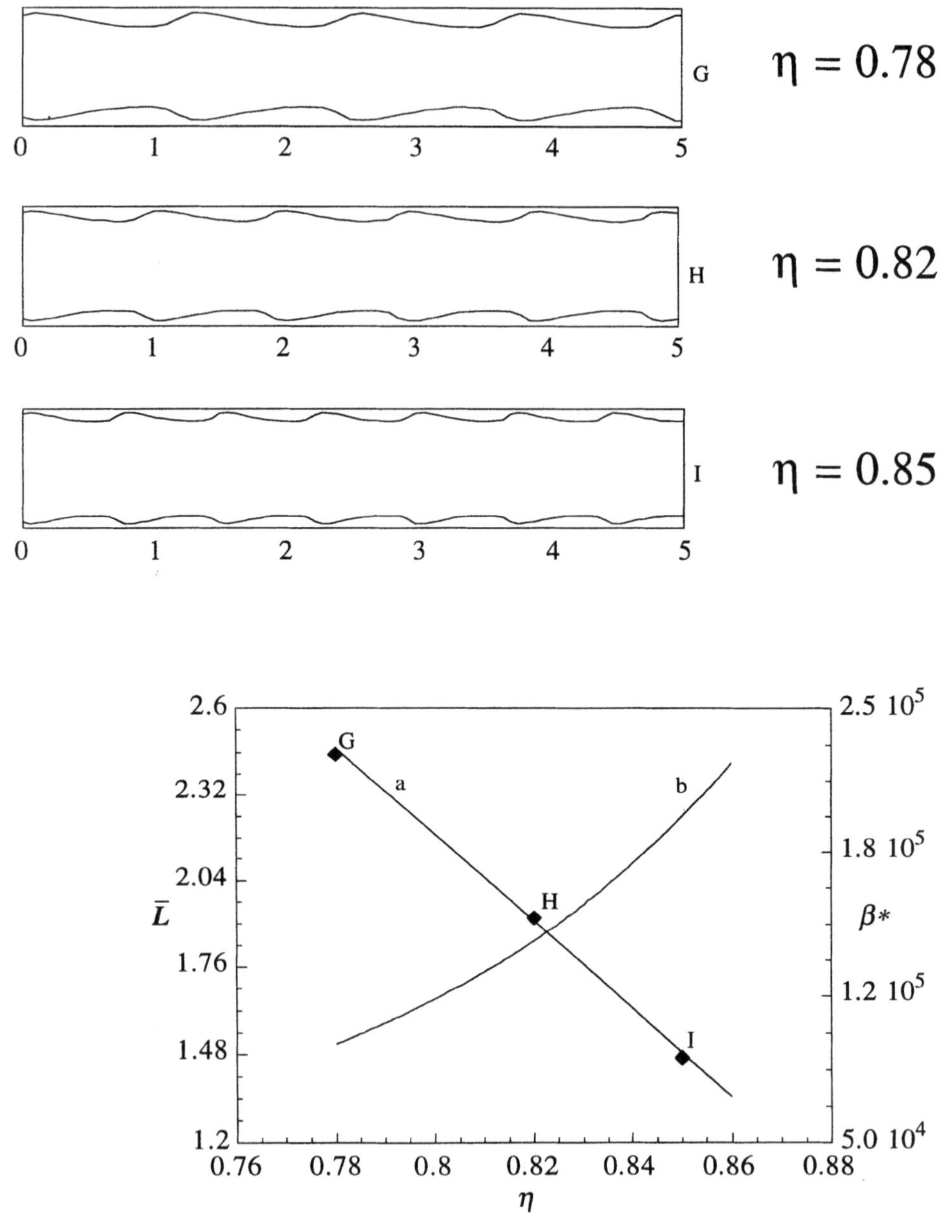

Figure 27. Numerical calculation of Bai, Kelkar and Joseph (1996). Wavelength $\overline{L} = 13.5 - 14.1\eta$ for $(\mathbb{R}, h) = (600, 1.4)$. **The wavelength and amplitude tend together to zero as $\eta \to 1$.** (see Joseph *et al.*, 1997)

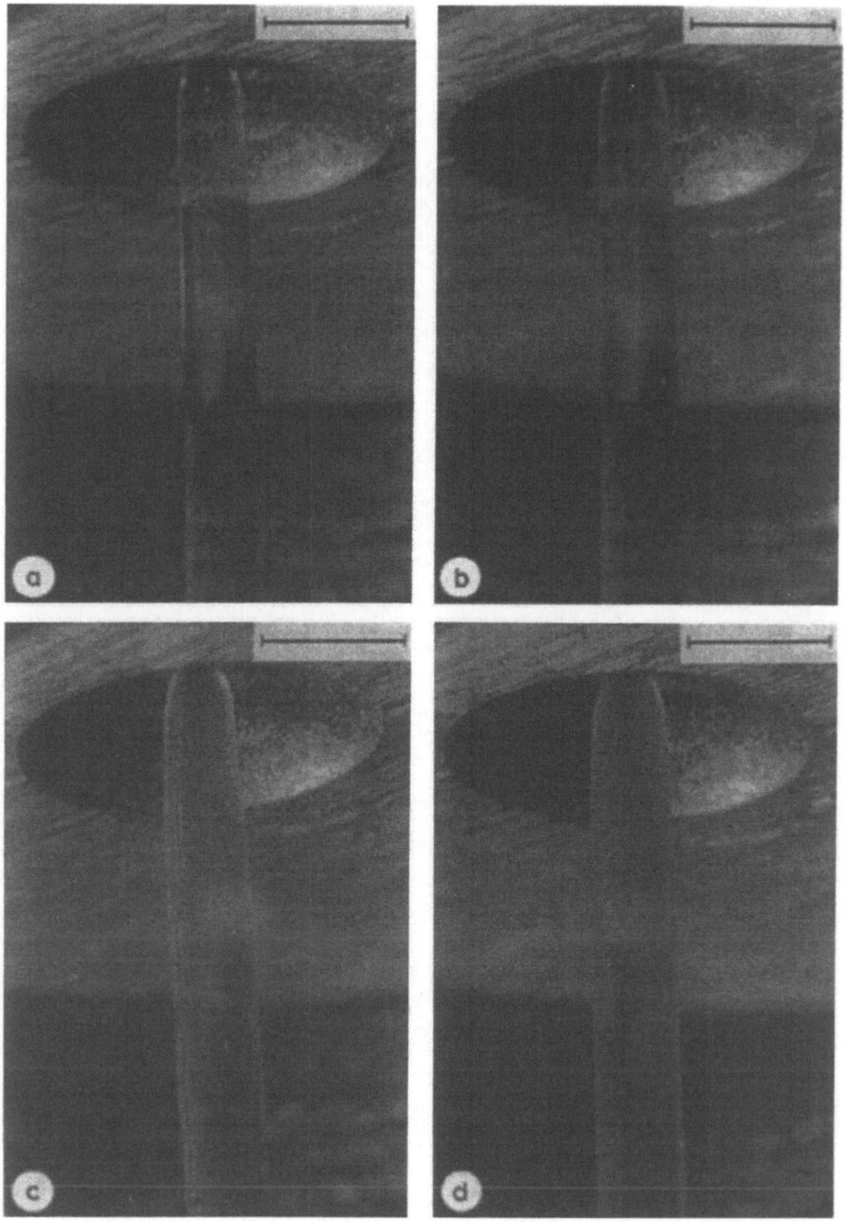

Figure 28. (Piau and Kissi, 1990) The short waves on (b), (c) and (d) are called "sharkskin". On them, and on the larger waves (c) through (b), the steep side advances.

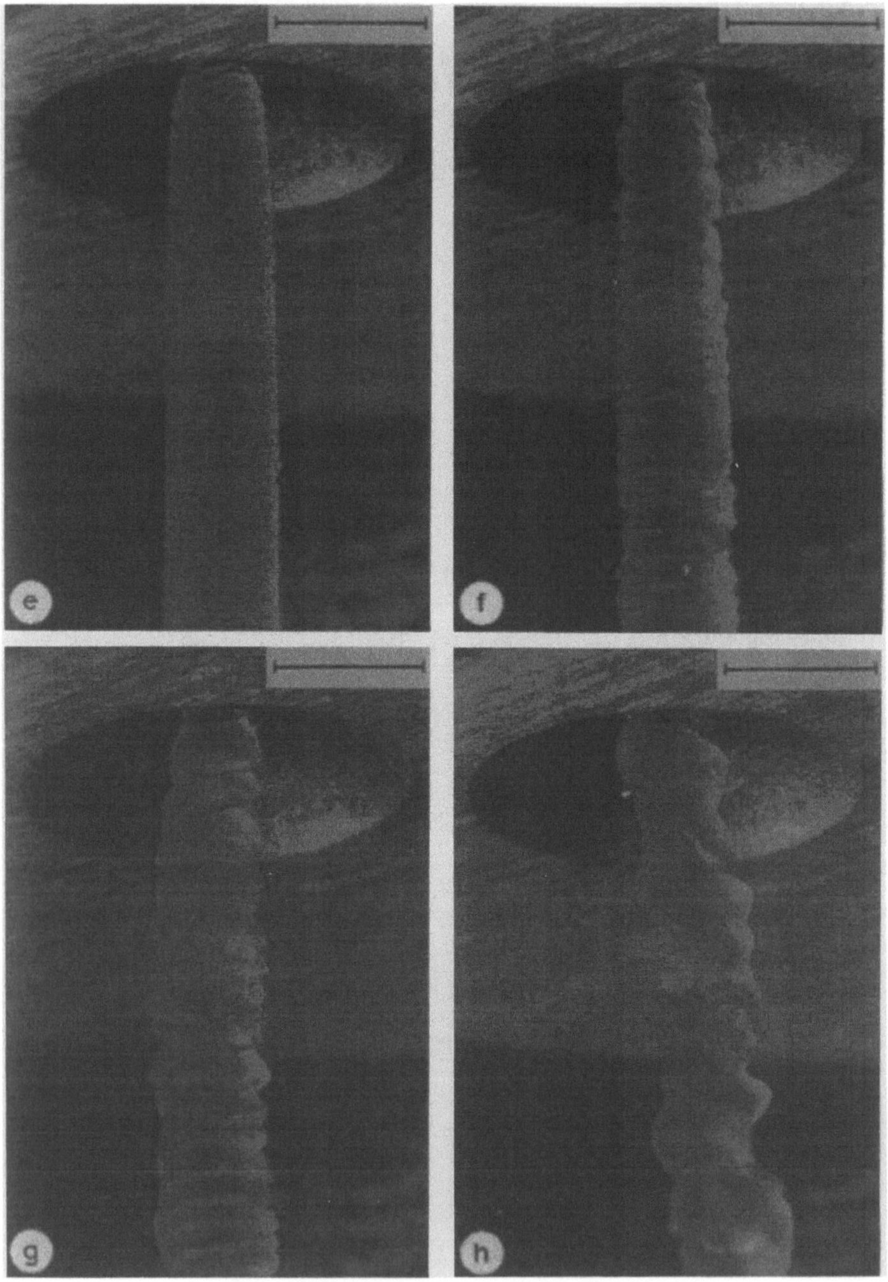

Figure 28, continued.

References

Arney M.S., Bai R., Guevara E., Joseph D.D., Liu K. (1993) Friction factor and holdup studies for lubricated pipelining, *Int. J. Multiphase Flow.*, **19** (6), 1061-1076.

Arney M.S., Ribeiro G.S., Guevara E., Bai R., Joseph D.D. (1995) Cement lined pipes for water lubricated transport of heavy oil, *Int. J. Multiphase Flow,* (In press).

Bai, R., Kelkar K., Joseph D.D. (1996) Direct simulation of interfacial waves in a high viscosity ratio and axisymmetric core annular flow, *J. Fluid Mech,* **327**, 1-34.

Brandt A., Bugliarello G. (1966) Concentration redistribution in the shear flow of monolayers of suspended particles. *Trans. Soc. Rheol.*, **10**(1), 229-251.

Charles M.E. (1963) The pipeline flow of capsules. Part 2: Theoretical analysis of the concentric flow of cylindrical forms, *Can. J. Chem. Engng.*, **46**.

Charles M.E., Govier G.W., Hodgson G.W. (1961) The horizontal pipeline flow of equal density of oil-water mixtures, *Can. J. Chem. Engng.* **39**, 17-36.

Chen, Y., Israelachvili, J., New mechanism of cavitation damage. *Science*, **252**, 1157-1160.

Feng J., Huang P.Y., Joseph D.D. (1995) Dynamic simulation of the motion of capsules in pipelines, *J. Fluid Mech.*, **286**, 201-207.

Huang A., Christodoulou C., Joseph D.D. (1994). Friction factor and holdup studies for lubricated pipelining. Part II: Laminar and k-e models of eccentric core flow, *Int. J. Multiphase Flow,* **20**(3), 481-91.

Issacs J.D., Speed J.B. (1904), U.S. Patent No. 759,374.

Iske P.L., Sergeev Y.A., Boersma W.H., Kurdyumov V.N., Moors J.H.J. (1995) Partical phase boundary layer theory of solid concentration and velocity profiles in circulating fluidized bed risers, in *Proceedings of the 2nd International Conference on Multiphase Flow* '95, Kyoto-Japan, FB1-33-37.

Joseph D.D. (1997) Steep wave fronts on extrudates of polymer melts and solutions: lubrication layers and boundary conditions, *JNNFM.*

Joseph D.D., Bannwart A., Liu J. (1996), Stability of annular flow and slugging, *Int. J. of Multiphase Flow,* **22**(6): 1247-54.

Joseph D.D., Bai R., Chen K.P., Renardy Y.Y. (1997) Core annular flows, **29**.

Joseph D.D., Liu J.Y. (1996) Steep wave fronts on extrudates of polymer melts and solutions. *J. Rheol.*

Joseph D.D., Nguyen K., Beavers G.S. (1986) Rollers, *Physics of Fluids* **29**, 2881.

Joseph D.D., Renardy Y.Y. (1993) Fundamentals of Two-Fluid Dynamics, New York: Springer-Verlag.

Kalika D.S., Denn M.M. (1987) Wall slip and extrudate distortion in linear low-density polyethylene, *J. Rheol.*, **31**, 815-834.

Kuhl, T., Ruths, M., Chen, Y.L., Israelachvili, J., Direct visualization of caviataion and damage in ultrathin liquid films. *The Journal of Heart Valve Disease*, **3** (suppl. I) 117-127 (1994).

Liu H. (1982) A theory of capsule lift-off in pipeline, *J. Pipelines 2*, 23-33.

Mandhane J.M., Gregory G.A., Aziz K. (1974) A flow pattern map for gas-liquid flow in horizontal pipes, *Int. J. Multiphase Flow*, **1**, 537-553.

Nez G.A., Briceo M., Mata C., Rivas H., Joseph D.D. (1996) Flow characteristics of concentrated emulsions of very viscous oil in water, *J. Rheol.*, (In press).

Ooms G., Segal A., Van der Wees A.J., Meerhoff R., Oliemans R.V.A. (1984) A theoretical model for core-annular flow of a very viscous oil core and a water annulus through a horizontal pipe. *Int. J. Multiphase Flow.* **10**, 41-60.

Piau J.M., Kissi N.E., Tremblay B., Influence of upstream instabilities and wall slip on melt fracture and sharkskin phenomena during silicones extrusions through orifice dies, *J. Non-Newtonian Fluid Mech.*, **34** (1990), 145-80.

Taitel Y., Dukler A.E. (1976), A model for predicting flow regime transitions in horizontal and near horizontal gas-liquid flow, *AIChE J.*, **22**, 47- 55.

EFFECTS OF HELICAL RIBS OF CAPSULE SURFACE ON A FLOW IN A HORIZONTAL HYDRAULIC PIPELINE

YOSHINARI FUJIWARA
Department of Mechanical Engineering, Kyushu Kyoritsu University
Jiyugaoka, Yahata-nishi, Kitakyushu 806 Japan
YUJI TOMITA
Department of Mechanical Engineering, Kyushu Institute of Technology
Sensuicho, Tobata, Kitakyushu 804 Japan

1. Introduction

It is well-known that in a hydraulic capsule pipeline capsules are smoothly transported without wheels due to lubricated water flow in the clearance between the capsule and the pipe wall. When wheels are attached to keep the capsule in concentric with the pipe, the mean velocity of capsule becomes higher than that of water, and this suggests that the core flow is favorable. To increase the capsule velocity and to reduce the pressure gradient due to capsules, Ellis *et al.* [1] argued that the clearance between the capsule and the pipe bottom should be increased and that the capsule should be supported in a slightly nose-up position. In order to raise the capsule from the pipe bottom they found that a simple collar around the leading half of the cylinder is the most effective. As for helical runners, they predicted in discussion [1] that helical runners would not increase the lift more than a collar, which raises the nose. They [2] also investigated several methods to reduce the pressure gradient by modification of the capsule surface which brings about the reduction of coefficient of friction.

In order to realize the core flow without using wheels, we experimentally investigate effects of helical ribs carved in a side surface of the capsule. The helical ribs deflect the flow and produce an extra pressure difference across the capsule. If the velocity of capsule is smaller than that of water, the pressure at the capsule front is decreased and the capsule velocity will be increased.

2. Concentric Annular Flow Model

As an idealized model of core flow for the capsule motion we take a concentric motion of a cylindrical capsule and assume that the capsule is kept in concentric with the pipe by wheels. The equation of motion of the capsule is given by

$$m\frac{d\dot{Z}}{dt} = F_P + F_S + F_C \tag{1}$$

where m is the capsule mass, $\dot{Z}$ is the capsule velocity, t is the time, F_P is the pressure force, F_S is the shear force and F_C is the solid frictional force. When the capsule planes in a pipeline without contact with the pipe wall and travels with constant velocity, we have

$$F_P + F_S = 0 \tag{2}$$

Thus, when F_P drives the capsule, F_S retards it. If we assume a fully developed laminar flow in the clearance between the capsule and the pipe wall, F_S is given by [3]

$$F_S = \pi dl\left\{\frac{kD}{8}\left(2+\frac{1-k^2}{k^2\ln k}\right)\frac{dp}{dx}+\frac{2\mu\dot{Z}}{kD\ln k}\right\} \tag{3}$$

where d is the capsule diameter, l is the capsule length, D is the pipe inside diameter, k is d/D and μ is the viscosity of water. dp/dx is the pressure gradient in the clearance and is given by

$$\frac{dp}{dx} = -\frac{V-\delta^2\dot{Z}}{\varepsilon^2} \tag{4}$$

where V is the water velocity. If $\dot{Z} > V/\delta^2$, both F_P and F_S become negative and this case happens in a decelerated motion. Therefore, for the equilibrium state of the capsule motion, we have

$$\dot{Z} \le V/\delta^2 \tag{5}$$

and $\dot{Z} \le 1.109V$ for $k = 0.9$, where

$$\varepsilon^2 = \frac{D^2(1-k^2)}{32\mu}\left(1+k^2+\frac{1-k^2}{\ln k}\right) \tag{6}$$

and

$$\delta^2 = -\frac{1-k^2}{2\ln k} \tag{7}$$

3. Experiment

Measurements were done under a constant tank head H from 0.5 to 2 mAq in a horizontal pipeline 30m in length and with an inside diameter of 40 mm. Figure 1 shows the ribbed

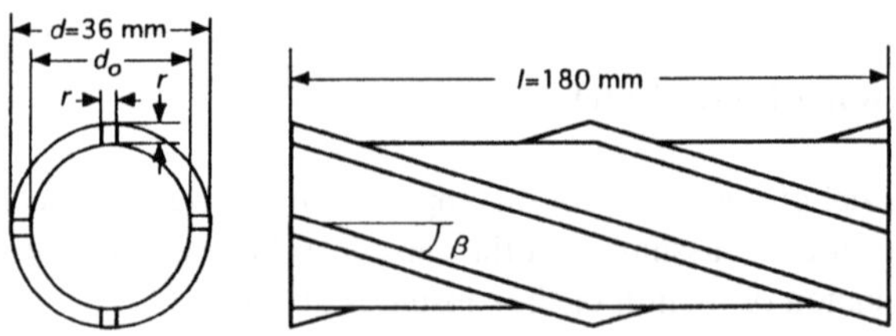

Figure 1. Geometry of ribbed capsule

TABLE 1. Specification of capsules

r(mm)	β(deg)	d_d(mm)	m(kg)	s
0		36	0.19	1.037
(wheeled)			0.33	1.801
3	0	30	0.19	1.421
	7.46			
	14.76		0.33	2.468
	21.44			
	27.64			
5	0	26	0.19	1.673
	6.47			
	12.78		0.33	2.906
	18.80			
	24.41			

capsule. Four ribs are carved in the side surface of the capsule whose length is 180 mm and outside diameter is 36 mm. The rib has a square cross section of which height r is 3 or 5 mm and the helical angle β is varied from 0 to 28 degrees, i.e., the turn angle of rib is varied from 0 to 360 degrees. TABLE 1 shows the specification of tested capsules and s is the density ratio of capsule to water. The wheeled capsules have four wheels at the front and the rear, respectively. The capsule mass was adjusted by using a weight. The motion of capsule was monitored by 38 photocells distributed along the pipeline. The line pressures and the pressure difference were measured by pressure transducers.

4. Results and Discussion

4.1. CAPSULE VELOCITY

The capsule velocity is obtained from the trajectory. The velocity of wheeled capsule is always higher than that of water and is close to $\dot{Z}/V = 1.109$ Figure 2 shows the velocity ratio of capsule to water $\dot{Z}/V$ against the helical angle for $r = 3$ mm and $r = 5$ mm ribs, respectively. From these figures it is seen that as a whole the capsule velocity is increased with an increasing helical angle and that the ribs are effective to increase the capsule velocity when the density ratio is large. It is observed that when the relative velocity between the capsule and the water is large, the revolution of the ribbed capsule is increased. However, when the velocity ratio is larger than unity, the helical ribs are not always effective and this is because, in turn, the flow behind the capsule is deflected.

4.2. LINE PRESSURE

A time history of line pressure was examined at 12.8 m from the pipe inlet during the

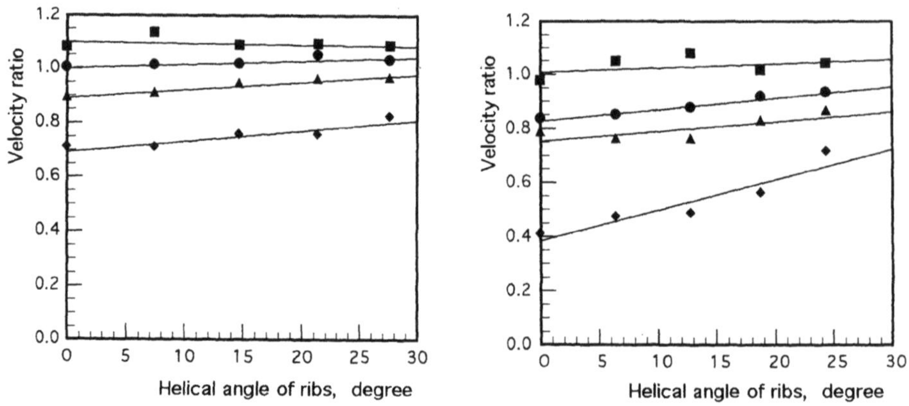

(a) $r = 3\,\mathrm{mm}$ (b) $r = 5\,\mathrm{mm}$

Figure 2. Velocity ratio of capsule to water: ■, $m = 0.19\,\mathrm{kg}$, $H = 2\,\mathrm{mAq}$; ●, $m = 0.19\,\mathrm{kg}$, $H = 1\,\mathrm{mAq}$; ▲, $m = 0.33\,\mathrm{kg}$, $H = 2\,\mathrm{mAq}$; ◆, $m = 0.33\,\mathrm{kg}$, $H = 1\,\mathrm{mAq}$.

capsule motion in the pipeline. For the wheeled capsule, there was not an appreciable pressure change in the history [4] and it was difficult to identify the time when the capsule passed the measuring point. This means that for the wheeled capsule the pressure drop in a region including the capsule is the same as that without the capsule. Figure 3 shows the results for the ribbed capsule. The capsule velocity is smaller than that of water in both cases. The time when the pressure steeply increases corresponds to the

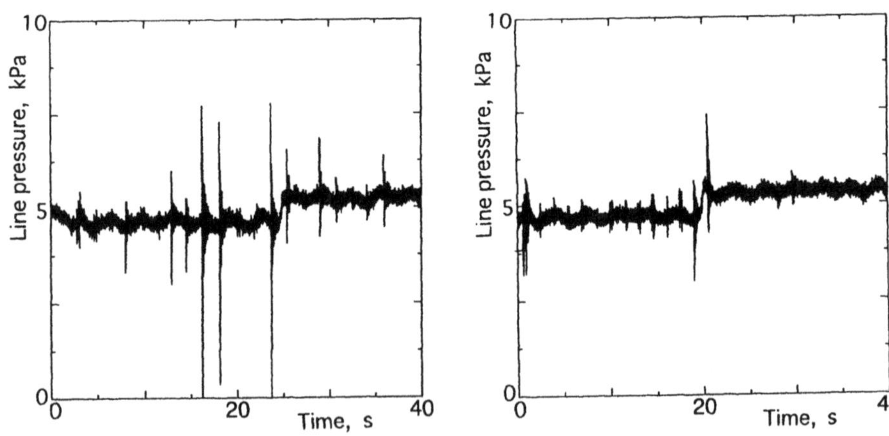

(a) $\beta = 0°$ (b) $\beta = 27.64°$

Figure 3. Time history of line pressure of ribbed capsule for $r = 3\,\mathrm{mm}$, $m = 0.33\,\mathrm{kg}$ and $H = 1\,\mathrm{mAq}$.

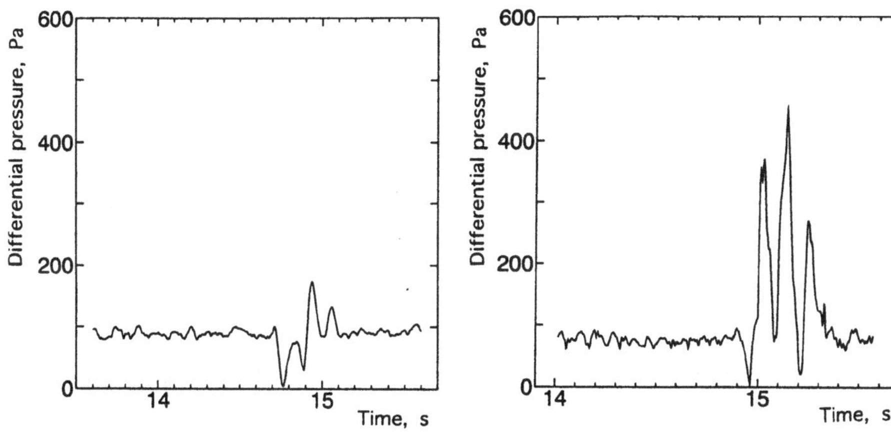

(a) m = 0.19 kg and H = 2 mAq (b) m = 0.33 kg and H = 1mAq

Figure 4. Differential pressure for wheeled capsule.

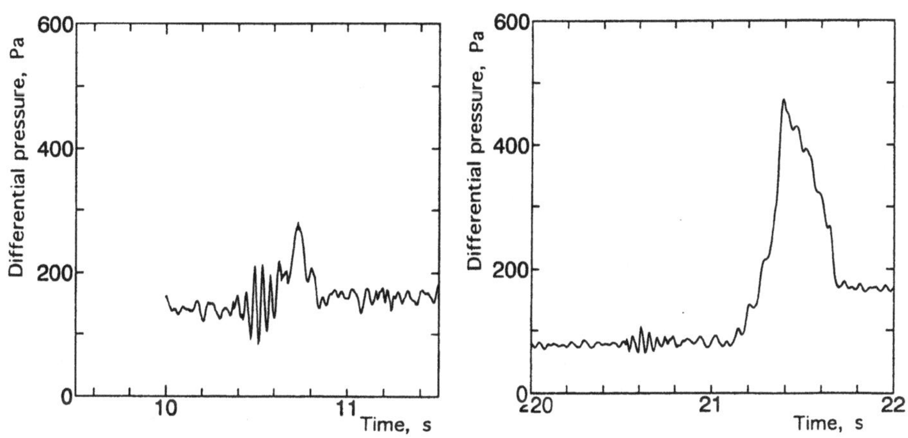

(a) r = 5mm , β = 24.41° , m = 0.19 kg and H = 2 mAq (b) r = 3 mm , β = 27.64° , m = 0.33 kg and H = 1mAq

Figure 5. Differential pressure for ribbed capsule.

capsule passing at the measuring point and the increase in line pressure means that the pressure drop in the region including the capsule is larger than that without the capsule or the pressure gradient behind the capsule is increased. It is found that when the helical angle is zero, that is, straight rib, there are pressure spikes of large amplitude and this suggests contacts of capsule with the pipe wall. However, if the helical angle is increased, such spikes disappear and this indicates that the contact of capsule with the pipe wall diminishes and that the capsule motion approaches the core flow. It was found that when the capsule velocity is larger than that of water there is not an appreciable pressure increase by the capsule passing.

4.3. DIFFERENTIAL PRESSURE

The differential pressure for 200 mm interval along the pipeline was measured a t 13.7 m from the pipe inlet. Figure 4 shows a time history of the differential pressure for the wheeled capsule. During the capsule passing the differential pressure fluctuates and the amplitude depends on the capsule mass. However, the pressure is substantially similar before and after the capsule passing. Figure 5 shows the result for the ribbed capsule. The velocity of capsule is larger than that of water in Figure 5 (a) and smaller than that in Figure 5 (b). It is found that there appears a beat-like wave before the capsule arrival and that the differential pressure is increased after the capsule passing when the velocity of capsule is smaller than that of water. This increase suggests that the pressure gradient behind the ribbed capsule is increased by the capsule motion. When the helical angle is decreased to zero, the result is similar to this except that the amplitude of pressure difference is increased. When in Figure 5 (a) the amplitude of beat-like wave is increased and the wave appears lately.

5. Conclusions

The helical ribs deflect the flow and produce an extra pressure difference. If the velocity of capsule is smaller than that of water, the increased pressure difference can contribute to drive the capsule. However, for the capsule which can move faster than water, the ribs are not always effective. Thus, the helical ribs are effective to increase the velocity of heavy capsules. The experimental results are summarized as follows:

1) The capsule velocity increases with an increasing helical angle of ribs. When water velocity is high and the capsule mass is small, the velocity of ribbed capsule becomes larger than that of water.

2) When the velocity of capsule is larger than that of water, the line pressure does not change by the capsule passing, while the capsule is faster than water the line pressure increases after the capsule passing.

3) The differential pressure at the pipeline is increased after the capsule passing when the velocity of ribbed capsule is smaller than that of water.

References

1. Ellis, H. S. and Kruyer, J. (1974) Minimizing the pressure gradients in capsule pipelines, *The Canadian J. of Chemical Engineering*, **52**, 457-462.

2. Ellis, H.S. and Kruyer, J. (1972) Some effects of fitting cylindrical pipeline capsule with collars or runners, *Hydrotransport* 2, paper F4 .

3. Tomita,Y., Yamamoto, M. and Funatsu, K. (1986) Motion of a single capsule in a hydraulic pipeline, *J. Fluid. Mech*. **171**, 495-508.

4. Tomita,Y, Okubo, T., Funatsu, K. and Fujiwara, Y. (1990) Unsteady Analysis of Hydraulic Capsule Transport in a Straight Horizontal Pipeline, in H. Liu and G.F. Round(eds.), *FREIGHT PIPELINES*, Hemisphere Publishing Corp., New York, pp.273-278.

NUMERICAL ANALYSIS OF DENSE PHASE LIQUID-SOLID FLOW IN A HORIZONTAL PIPE

S.HARADA, T.TANAKA AND Y.TSUJI
Department of Mechanical Engineering, Osaka University
2-1, Yamada-oka, Suita, Osaka, 565, JAPAN

AND

K.ASAKURA
Department of Mineral Resources Development Engineering, Hokkaido University
N13 W8, Sapporo, Hokkaido, 060, JAPAN

Abstract. Numerical simulations of the liquid-solid flow in a horizontal pipe were performed by the Lagrangian/Eulerian approach. Motion of individual coarse particles was calculated by the Newton's equation of motion. The contact forces were given by DEM (Discrete Element Method)[1] model, which enables ones to estimate the forces on the moving particles in contact with their neighbors. The motion of liquid was calculated by solving the local averaged equations of continuity and motion. The calculated results showed the characteristic flow patterns of liquid-solid systems, such as flows with a moving bed or a deposit. The relation between the fluid velocity and the flow pattern agreed with the experimental results of sediment transport.

1. Introduction

When the transport velocity is small in the hydraulic conveying of solid materials, the solid particles form a deposit on the bottom of the pipe. In such a condition, the fluid motion is affected largely by the motion of the large-scale structure of particle assemblies, not by individual particles.

As an approach to the dense system, the macroscopic analyses such as the layer models have been developed and used to estimate properties of hydraulic transport such as pressure loss. However, this approach requires empirical assumptions of the velocity and the stress distributions within the bed, so that more microscopic analysis based on particle dynamics which require less assumptions is desired. Recently, the advance in computer performance has made it possible to make numerical simulations of liquid-solid flows based on basic dynamics. For example, Hu[2] simulated directly the coupled motion of liquid and particles using the finite element method with the moving mesh and showed the structure of liquid-solid flow at low Reynolds number. Such microscopic analyses can predict the fluid motion in detail without any empirical models, but the number of the particles treated is limited.

In order to describe the macroscopic motions of fluid-solid systems, basic equations that consist of local mean variables have been developed[3]. Tsuo *et al.*[4] simulated the characteristic motion in the gas-solid fluidized bed using such averaged equation models. Porosity distribution obtained by them agreed with the measurement. However, since the solid phase is assumed as a continuum, this method is suitable only for the systems which involve small particles compared with the scale of the system.

DEM (Discrete Element Method) is the only method for numerical simulation of motion of many particles in successive contacts with other particles or walls. It has been applied to analyses of various granular systems. For example, Tsuji *et al.*[5] simulated flows in a two-dimensional gas-solid fluidized bed by applying the local averaged model to the fluid and DEM to the particles. They calculated trajectories of individual particles and the obtained macroscopic structures such as bubble formation or circulation in the particle bed which are characteristic features of gas fluidized beds. However, this method has been rarely used in liquid-solid flow systems.

In this study, DEM was applied to flow of liquid-solid mixture. The aim of this study is to understand the detailed flow structure in hydrotransport systems. Three dimensional analysis was performed on both phases to obtain the flow pattern in the practical conditions. The distribution of solid stress and the deformation of particle layer can be estimated by the simulation results and are applicable to the modeling of constitutive equations in continuum models.

2. Formulation and Modeling

2.1. CONTACT FORCE MODEL IN DEM

The inter-particle collision in liquid has been studied in the field of elasto-hydrodynamics, which deals with coupling problems of elasticity and hydrodynamics. Recently, some contact models taking account of the elastic deformation of particles were proposed[6][7]. However, it is not easy to extend these models to multi-body problems exactly. According to these studies, if the inertia of a particle is large compared with the viscous force, the restitution coefficient is independent of the initial approach velocity[6]. Therefore, in the present work, we used the contact force model which gives constant restitution coefficient[8] since the particle velocity is relatively high in the present condition.

In DEM, the contact force on a particle is expressed by using springs, dashpots and a friction slider as shown in Fig. 1. In the case of spherical particles, a non-linear spring should be used for modeling the contact force in the normal direction. The spring coefficient is determined from the physical properties of the particle material and the curvature of the particle surfaces at the contact point. Similarly, the tangential stiffness is expressed by the properties of the particles[8]. If the static friction exceeds a limit, the dynamic friction acts on the particles by the friction slider.

If the time step is chosen sufficiently small, the instantaneous contact force exerted on each particle is determined by the instantaneous relative motion. Therefore the motion of each particle can be found by integrating the equation of motion with the contact force between immediate neighbors.

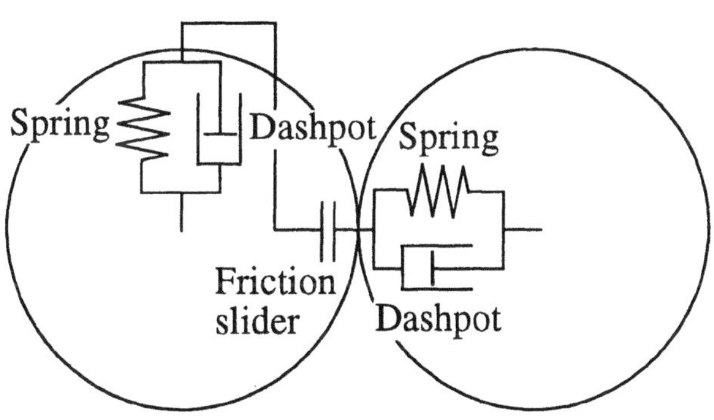

Figure 1. Contact force model

2.2. GOVERNING EQUATIONS

The motion of a particle is calculated by the Newton's equation of motion which includes the fluid force $\boldsymbol{F}_f$ and the contact force $\boldsymbol{F}_c$,

$$\ddot{\boldsymbol{x}} = \frac{1}{m_p}(\boldsymbol{F}_f + \boldsymbol{F}_c) + \boldsymbol{g}, \tag{1}$$

where m_p is the mass of the particle and $\boldsymbol{g}$ is the gravitational acceleration. The fluid force $\boldsymbol{F}_f$ is the sum of the drag force, the Saffman's lift force[9], the Magnus force and the forces caused by the pressure gradient and by the virtual added mass. The equation of rotational motion for the particle consists of the viscous resistant torque $\boldsymbol{T}_v$ and the torque caused by the contact force $\boldsymbol{T}_c$,

$$\dot{\boldsymbol{\omega}} = \frac{1}{I_p}(\boldsymbol{T}_v + \boldsymbol{T}_c), \tag{2}$$

where I_p is the moment of particle inertia.

Basic equations for fluid phase are the equations of continuity and motion based on the local averaged theory written in cylindrical coordinates,

$$\frac{\partial \epsilon_f}{\partial t} + \frac{1}{r}\frac{\partial (r\epsilon_f u_r)}{\partial r} + \frac{1}{r}\frac{\partial (\epsilon_f u_\theta)}{\partial \theta} + \frac{\partial (\epsilon_f u_z)}{\partial z} = 0, \tag{3}$$

$$\begin{aligned}
&\frac{\partial (\epsilon_f u_r)}{\partial t} + \frac{1}{r}\frac{\partial (r\epsilon_f u_r u_r)}{\partial r} + \frac{1}{r}\frac{\partial (\epsilon_f u_r u_\theta)}{\partial \theta} + \frac{\partial (\epsilon_f u_r u_z)}{\partial z} = -\frac{\epsilon_f}{\rho_f}\frac{\partial p}{\partial r} + \frac{\epsilon_f u_\theta^2}{r} \\
&+\frac{\nu}{r}\frac{\partial}{\partial r}\left(2r\epsilon_f\frac{\partial u_r}{\partial r}\right) + \frac{\nu}{r}\frac{\partial}{\partial \theta}\left[\epsilon_f\left(\frac{1}{r}\frac{\partial u_r}{\partial \theta} + \frac{\partial u_\theta}{\partial r} - \frac{u_\theta}{r}\right)\right] \\
&+\nu\frac{\partial}{\partial z}\left[\epsilon_f\left(\frac{\partial u_r}{\partial z} + \frac{\partial u_z}{\partial r}\right)\right] - \frac{2\epsilon_f \nu}{r}\left(\frac{1}{r}\frac{\partial u_\theta}{\partial \theta} + \frac{u_r}{r}\right) + \frac{f_{pr}}{\rho_f} + \epsilon_f g_r,
\end{aligned} \tag{4}$$

$$\begin{aligned}
&\frac{\partial (\epsilon_f u_\theta)}{\partial t} + \frac{1}{r}\frac{\partial (r\epsilon_f u_r u_\theta)}{\partial r} + \frac{1}{r}\frac{\partial (\epsilon_f u_\theta u_\theta)}{\partial \theta} + \frac{\partial (\epsilon_f u_\theta u_z)}{\partial z} = -\frac{\epsilon_f}{\rho_f r}\frac{\partial p}{\partial \theta} - \frac{\epsilon_f u_r u_\theta}{r} \\
&+\frac{\nu}{r}\frac{\partial}{\partial r}\left[r\epsilon_f\left(\frac{\partial u_\theta}{\partial r} + \frac{1}{r}\frac{\partial u_r}{\partial \theta} - \frac{u_\theta}{r}\right)\right] + \frac{\nu}{r}\frac{\partial}{\partial \theta}\left[2\epsilon_f\left(\frac{1}{r}\frac{\partial u_\theta}{\partial \theta} + \frac{u_r}{r}\right)\right] \\
&+\nu\frac{\partial}{\partial z}\left[\epsilon_f\left(\frac{\partial u_\theta}{\partial z} + \frac{1}{r}\frac{\partial u_z}{\partial \theta}\right)\right] + \frac{\epsilon_f \nu}{r}\left(\frac{\partial u_\theta}{\partial r} + \frac{1}{r}\frac{\partial u_r}{\partial \theta} + \frac{u_\theta}{r}\right) + \frac{f_{p\theta}}{\rho_f} + \epsilon_f g_\theta,
\end{aligned} \tag{5}$$

$$\begin{aligned}
&\frac{\partial (\epsilon_f u_z)}{\partial t} + \frac{1}{r}\frac{\partial (r\epsilon_f u_r u_z)}{\partial r} + \frac{1}{r}\frac{\partial (\epsilon_f u_\theta u_z)}{\partial \theta} + \frac{\partial (\epsilon_f u_z u_z)}{\partial z} = -\frac{\epsilon_f}{\rho_f}\left(\frac{\partial p}{\partial z} - b\right) \\
&+\frac{\nu}{r}\frac{\partial}{\partial r}\left[r\epsilon_f\left(\frac{\partial u_z}{\partial r} + \frac{\partial u_r}{\partial z}\right)\right] + \frac{\nu}{r}\frac{\partial}{\partial \theta}\left[\epsilon_f\left(\frac{1}{r}\frac{\partial u_z}{\partial \theta} + \frac{\partial u_\theta}{\partial z}\right)\right] \\
&+\nu\frac{\partial}{\partial z}\left[2\epsilon_f\frac{\partial u_z}{\partial z}\right] + \frac{f_{pz}}{\rho_f} + \epsilon_f g_z,
\end{aligned} \tag{6}$$

where ϵ_f is the local volume fraction of the fluid and b in Eq.(6) is the mean pressure gradient, mentioned in Section 3. Eqs.(3)~(6) do not describe the microscale motion in the gap between particles, but the large-scale motion of fluid flow. The influence of the solid phase is expressed by the reaction from the particle f_p and the volume fraction of the fluid ϵ_f.

2.3. MODELING OF INTERACTION

When the solid motion is calculated discretely like this work, it is necessary to model the fluid forces on the particles which should be consistent with the continuum model. In the case of the Eulerian/Eulerian method[10], both phases are assumed as continua and the macroscopic equations of motion per unit volume for each phase are as follows,

$$\frac{\partial}{\partial t}\rho_f\epsilon_f\boldsymbol{u}_f + \nabla\cdot\rho_f\epsilon_f\boldsymbol{u}_f\boldsymbol{u}_f = -\epsilon_f\nabla p - \beta(\boldsymbol{u}_f - \boldsymbol{u}_s) + \nabla\cdot\epsilon_f\boldsymbol{\tau}_f + \rho_f\epsilon_f\boldsymbol{g}, \quad (7)$$

$$\frac{\partial}{\partial t}\rho_s\epsilon_s\boldsymbol{u}_s + \nabla\cdot\rho_s\epsilon_s\boldsymbol{u}_s\boldsymbol{u}_s = -\epsilon_s\nabla p + \beta(\boldsymbol{u}_f - \boldsymbol{u}_s) + \nabla\cdot\epsilon_s\boldsymbol{\tau}_s + \rho_s\epsilon_s\boldsymbol{g}, \quad (8)$$

where $\boldsymbol{\tau}$ is the stress tensor and ϵ_s is the volume fraction of solid phase, which satisfies $\epsilon_f + \epsilon_s = 1$. The subscript s indicates the solid phase as a continuum. Eq.(7) corresponds to Eqs.(4)~(6).

β in Eqs.(7) and (8) is the coefficient of interphase friction and is calculated by means of slip velocity between both phases,

$$\epsilon_f \leq 0.8, \quad \beta = 150\frac{\epsilon_s^2}{\epsilon_f}\frac{\mu}{d_p^2} + 1.75\epsilon_s\frac{\rho_f|\boldsymbol{u}_f - \overline{\boldsymbol{u}_p}|}{d_p}, \quad (9)$$

$$\epsilon_f > 0.8, \quad \beta = \frac{3}{4}C_D\rho_f\frac{\epsilon_s}{\epsilon_f^{1.7}}\frac{|\boldsymbol{u}_f - \overline{\boldsymbol{u}_p}|}{d_p}, \quad (10)$$

where d_p is the diameter of the particle, C_D is the drag coefficient and $\overline{\boldsymbol{u}_p}$ is the mean particle velocity per unit volume. The subscript p indicates the discrete particle. The Ergun's equation revised by local superficial velocity is used for the dense region. For the dilute region, the standard drag force corrected empirically by means of the fluid volume fraction is used.

In the case that the solid motion is calculated discretely, the force integrated by unit volume V in Eq.(8) should be counterbalanced with the sum of the forces exerted on each particle,

$$- \iiint_V \epsilon_s \nabla p dv + \iiint_V \beta(\boldsymbol{u}_f - \boldsymbol{u}_s) dv + \iiint_V \nabla \cdot \epsilon_s \boldsymbol{\tau}_s dv + \iiint_V \rho_s \epsilon_s \boldsymbol{g} dv$$
$$\simeq - \sum^N \nabla p V_p + \sum^N \frac{1}{\epsilon_s} \beta(\boldsymbol{u}_f - \boldsymbol{u}_p) V_p + \sum^N \boldsymbol{F}_c + \sum^N \rho_s \boldsymbol{g} V_p, \quad (11)$$

where N is the number of the particles per unit volume and V_p is the volume of one particle. The equation of motion for each particle is derived from Eq.(11). That is,

$$m_p \frac{d\boldsymbol{u}_p}{dt} = - \nabla p V_p + \frac{1}{\epsilon_s} \beta(\boldsymbol{u}_f - \boldsymbol{u}_p) V_p + \boldsymbol{F}_c + \rho_s V_p \boldsymbol{g}$$
$$= - \nabla p V_p + \frac{V}{N} \beta(\boldsymbol{u}_f - \boldsymbol{u}_p) + \boldsymbol{F}_c + \rho_s V_p \boldsymbol{g}. \quad (12)$$

Hence, the pressure force and the drag force acting on the particle are given as the first and the second term in Eq.(12) respectively. In fact, other components of the fluid force such as the lift force act on the particle. However, the reactions of these forces to the fluid phase are not considered in the present work, because these forces hardly affect the macroscopic fluid motion.

3. Calculation Conditions

3.1. COORDINATES AND BOUNDARIES

The calculation domain is the inside of a horizontal cylinder with D=42.5mm in diameter and L=500mm in length. The boundaries in the longitudinal direction are periodic in both phases. The mean pressure gradient b in Eq.(6) acts as the driving force of the fluid flow and is adjusted so as to keep the mean flow rate constant.

The motions of the particles are described in the Cartesian coordinates, while the fluid motion is described in the cylindrical ones. Therefore the coordinates for some quantities have to be converted from the one to the other for evaluating interphase friction.

3.2. PHYSICAL PROPERTIES

The physical properties of the particles and the fluid used in the present simulation are shown in Table 1. The properties of the pipe wall concerning the particle-wall contact are assumed to be those of acrylic wall. The friction coefficient of the particles in water was set equal to 0.26.

TABLE 1. Physical properties of solid-liquid mixtures

Particle diameter d_p[mm]	Particle density ρ_s[kg/m^3]	Young's modulus E_p[N/m^2]	Poisson's ratio σ_p[-]	Fluid density ρ_f[kg/m^3]	Fluid viscosity ν_f[kg/m·s]
2.07	2480	1.0×10^5	0.33	1000	1.002×10^{-3}

3.3. DECISION OF TIME STEP

The time step for calculating the particle motion must be sufficiently smaller than the time scale of particle collision. The critical time step in DEM depends on the particle mass and the spring stiffness, and can be estimated roughly by single degree-of-freedom system[1].

On the other hand, the calculation of fluid flow requires the time step which is sufficiently smaller than the time scale of a change in the flow. Especially, the method of calculation of the local volume fraction largely affects the critical time step for the fluid motion.

Calculation of the particle volume fraction in a calculation cell is sensitive to where the particle locates relative to the calculation cell. Particularly this is the case when the particle size is comparatively large. In this situation, a small time step sometimes leads to instability of flow calculation through the time derivative of the volume fraction. Taking account of the above problem, the time step Δt for the both phases was set equal to 1.0×10^{-4} sec. in this simulation.

4. Computed Results

It is known that there are four flow regimes for liquid-solid flows in a horizontal pipe. Fig. 2 shows the diagram of flow regime plotted in the frame of the mean fluid velocity and the particle diameter for pipe diameter D=42.5mm. This diagram is drawn by well-known empirical relations of hydraulic transport[11]. This figure shows that a deposition builds up at a velocity below the critical velocity U_c, and as the flow velocity increases, the particle flow pattern changes into more dispersed one.

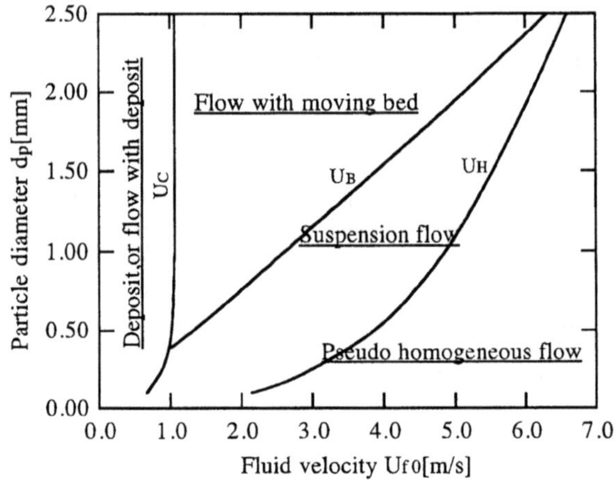

Figure 2. Flow regimes and three transition velocities for pipe diameter D=42.5mm

4.1. UNSTEADY MOTION

In order to study the flow in the sediment regimes shown in Fig. 2, numerical simulations were done in the conditions around the critical velocity U_c. Fig. 3(a) shows the particle distribution in the initial condition, where the particles are in the whole pipe cross section but in the half length of the pipe. The positions of individual particles are randomly. The distribution of the fluid velocity in the initial condition was assumed to be that of the fully developed laminar flow. The mean fluid velocity is U_{f0}=0.87 m/s. The number of the particles is 13191 which corresponds to the particle volume fraction 8.64%.

Snapshots of the particles are shown in Fig. 3. The particles accelerate due to the fluid force, settling and forming a moving bed on the bottom. The leading particles catch up with the tail of the bed through the periodic boundaries and flow onto it. Finally, they form a stationary bed near the pipe wall. The axial fluid velocity profile at t=0.40 sec. corresponding to Fig. 3(b) is shown in Fig. 4. It is found that the fluid flow develops in the free-flow regime, and that the particle bed has large effects on the profile.

Fig. 5 shows the cross-sectional fluid velocity at z/L=0.60. In the early stage, the descending motion of the particles causes an ascending flow of the liquid. Subsequently a pair of circulation is formed as shown in Fig.

5(a),(b). After the establishment of the stationary bed, the circulation pair are remained(Fig. 5(c)) within the range of present calculation.

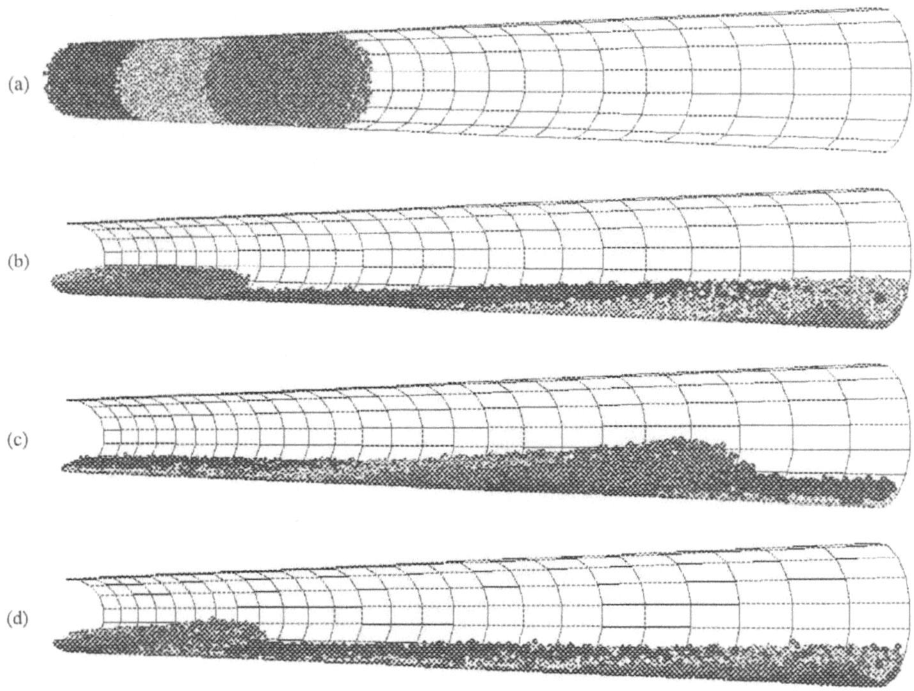

Figure 3. Snapshots of the particles for U_{f0}=0.87m/s at (a) t=0, (b) t=0.40, (c) t=0.80 and (d) t=1.20 sec.

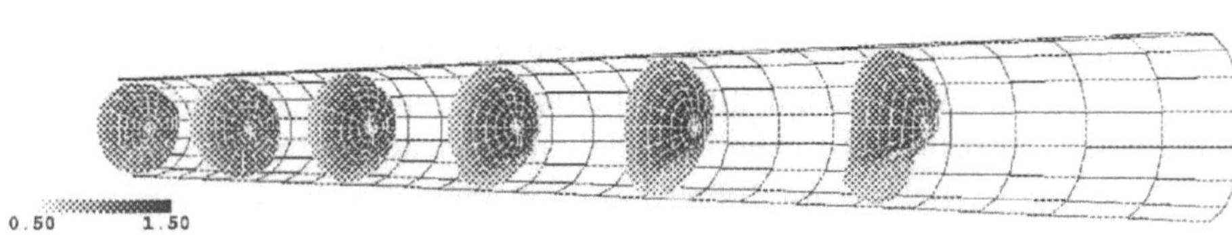

Figure 4. Fluid velocity profile in longitudinal direction for U_{f0}=0.87m/s at t=0.40 sec.

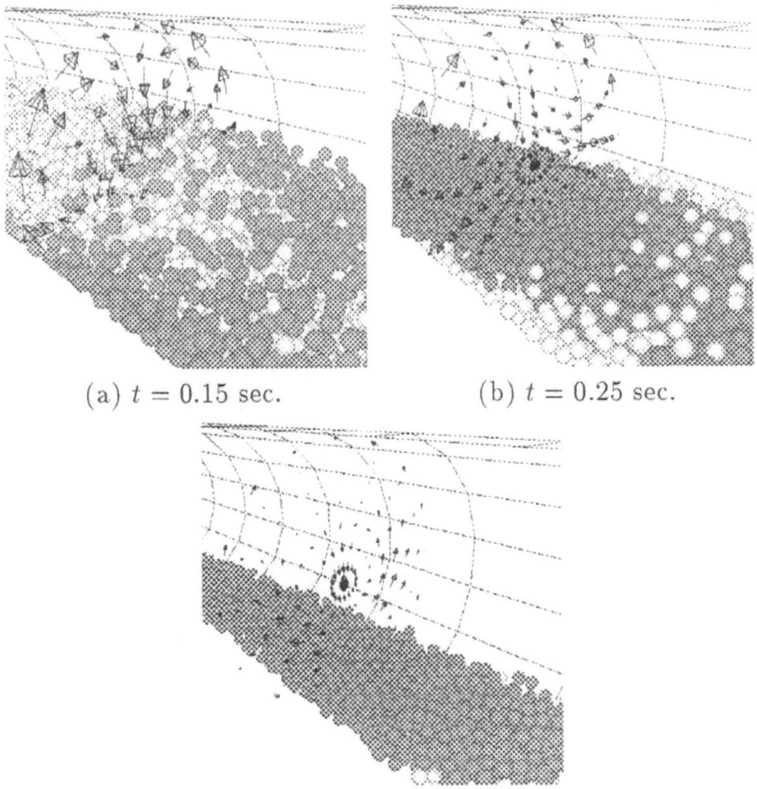

(a) $t = 0.15$ sec.

(b) $t = 0.25$ sec.

(c) $t = 0.40$ sec.

Figure 5. Instantaneous fluid velocity vectors in a cross-section

4.2. STABLE FLOW PATTERN

Comparison of the flow patterns between two different velocities around the critical velocity U_c was made. As an initial condition, the stationary bed was formed by the particles piled randomly on the pipe, and the Poiseuille flow was given as an initial fluid velocity. The number of the particle is 9350 which corresponds to the particle volume fraction 6.24%.

First, the particles were diffused and suspended because of large initial fluid force. Subsequently they settled down on the bottom of the pipe gradually and form a particle bed. Fig. 6 shows the snapshots of the particles at $t = 1.20$ sec. and Fig. 7 is the corresponding mean particle velocity distribution on the vertical center plane. In the case of $U_{f0} = 0.87$m/s, the particle bed form a deposit. It is observed that some particles tumble on the bed surface.

On the other hand, in the case of U_{f0}=1.84m/s, a moving bed as shown in these figures is formed. Some particles are suspended. The relation between the mean fluid velocity U_{f0} and the flow pattern is consistent with the empirical relation in Fig. 2.

This behavior of particles seems to be tranquil compared with the actual flow patterns. This is because the fluid turbulence is neglected in the present simulation.

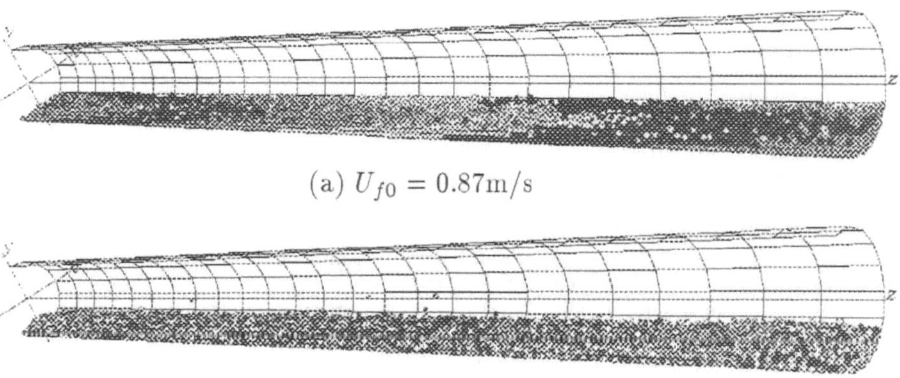

(a) $U_{f0} = 0.87$m/s

(b) $U_{f0} = 1.84$m/s

Figure 6. Snapshots of the particles for $U_{f0} = 0.87$m/s and U_{f0}=1.84m/s at t=1.20 sec.

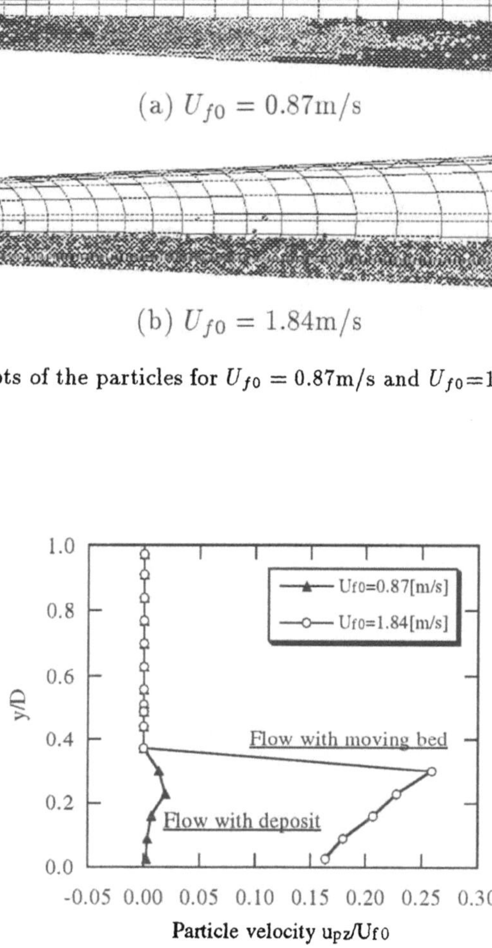

Figure 7. Mean particle velocity at vertical center plane for U_{f0}=0.87m/s and 1.84m/s

5. Conclusions

Three dimensional simulations of liquid-solid flows of coarse particles in a horizontal pipe were performed. The Lagrangian/Eulerian approach was used and DEM was applied to the calculation of particle motion. The obtained results showed the characteristic flow patterns of particles, such as flows with a moving bed or a deposit. In addition, the present relation between the mean fluid velocity and the flow pattern is consistent with the diagram of flow regimes predicted by previous works.

References

1. Cundall, P.A. and Strack, O.D.L. (1979) A discrete numerical model for granular assembles, *Geotechnique*, **29**, 47–65.
2. Hu, H.H. (1996) Direct simulation of flows of solid-liquid mixtures, *Int. J. Multiphase Flow*, **22**, 335–352.
3. Anderson, T.B. and Jackson, R. (1967) A fluid mechanical description of fluidized beds, *I&EC Fundamentals*, **6**, 527–539.
4. Tsuo, Y.P. and Gidaspow, D. (1990) Computation of flow patterns in circulating fluidized beds, *AIChE Journal*, **36**, 885–896.
5. Tsuji, Y., Kawaguchi, T. and Tanaka, T. (1993) Discrete particle simulation of two-dimensional fluidized bed, *Powder Technology*, **77**, 79–87.
6. Davis, R.H., Serayssol, J. and Hinch, E.J. (1986) The elastohydrodynamic collision of two spheres, *J. Fluid Mech.*, **163**, 479–497.
7. Lian, G., Adams, M.J. and Thornton, C. (1996) Elastohydrodynamic collisions of solid spheres, *J. Fluid Mech.*, **311**, 141–152.
8. Tsuji, Y., Tanaka, T. and Ishida, T. (1992) Lagrangian numerical simulation of plug flow of cohesionless particles in a horizontal pipe, *Powder Technology*, **71**, 239–250.
9. Saffman, P.G. (1965) The lift on a small sphere in a slow share flow, *J. Fluid Mech.*, **22**, 385–400.
10. Bouillard, J.X., Lyczkowski, R.W. and Gidaspow, D. (1989) Porosity distributions in a fluidized bed with an immersed obstacle, *AIChE Journal*, **35**, 908–922.
11. Graf, W.H. (1971) *Hydraulics of Sediment Transport*, McGraw-Hill Book Company

PARAFFIN DEPOSITION PHENOMENA IN CRUDE OIL PIPELINES

James P. Brill
Petroleum Engineering Dept.
The University of Tulsa

1. Introduction

When oil and gas are produced from wells drilled in deep water environments, the fluids undergo rapid temperature reductions as a result of heat transfer to the surrounding cold water. Typical sea floor temperatures in 2,000-m water depths are often about 5 °C. When the oil temperature near the pipe wall cools below the cloud point or wax appearance temperature, deposition of paraffin on the pipe wall begins. Predicting the phenomena of paraffin deposition requires combining computations for heat transfer, phase equilibrium thermodynamics, fluid dynamics and deposition mechanisms, some of which are not well understood for either single phase or multiphase flow conditions. Once started, paraffin deposition can continue until significant reductions occur in the pipe cross-sectional area available for flow, sometimes resulting in total blockage.

A new 4-year, $4 million Joint Industrial Project (JIP) was initiated in 1995 at The University of Tulsa to investigate wax deposition in multiphase flowlines and wellbores. This paper presents part of Phase 1 of the JIP, which has the objective of investigating, experimentally and theoretically, paraffin deposition during single phase oil flow in a near-horizontal flowline (Matzain, 1997). A new test facility was constructed and then used to gather preliminary wax deposition data as the basis to verify that the test facility could deposit wax, to establish the desired operating conditions, and to identify what modifications must be made to the facility to gather additional high quality single phase deposition data. Following sections review selected previous publications on the subject of paraffin deposition, describe the test facility constructed, and present trends in preliminary deposition data obtained with the facility.

2. Literature Review

The precipitation of paraffin solids or wax normally occurs when the temperature of the pipe wall falls below the hydrocarbon liquid cloud point temperature. Cloud point temperature is defined as the temperature at which wax particles first begin to precipitate from solution. Wax could also form near the pipe wall even if the bulk oil is at a temperature above its cloud point. Cole *et al.* (1960) claimed that the temperature differential between the solution cloud point and the cold surface is more important than the differential between the bulk oil and the cold surface. Deposition will occur when the inner pipe wall surface temperature is below both the solution temperature and the solution cloud point.

Wax deposition in a pipeline is probably governed by two mechanisms: molecular diffusion of paraffin molecules dissolved in solution; and, shear dispersion. Gravity settling is not a significant mechanism due to the small difference in density between the wax and oil. Differential temperature effects control the diffusion mechanism, while flow rate controls the shear dispersion mechanism. Bern *et al.* (1980) illustrated that molecular diffusion transports wax which is in solution, whereas shear dispersion transports the particulate wax.

2.1 MOLECULAR DIFFUSION

The rate of transport of dissolved wax through the molecular diffusion mechanism is given by Fick's diffusion law in Equation 1, where the radial concentration gradient is expressed in terms of a temperature concentration gradient and a radial temperature gradient.

$$\begin{aligned}\frac{dm_m}{dt} &= \rho_w D_m A_i \frac{dC_w}{dr} \\ &= \rho_w D_m A_i \frac{dC_w}{dT}\frac{dT}{dr} \qquad (1)\end{aligned}$$

It should be possible to predict the temperature concentration gradient using appropriate thermodynamic models for predicting soild-liquid-vapor equilibrium behavior for multicomponent hydrocarbon systems. The radial temperature gradient can be calculated with appropriate heat transfer models.

2.2 SHEAR DISPERSION

Burger *et al.* (1981) reported that the shearing of fluid close to the pipe wall induces a lateral movement of particles from a region of higher velocity towards a region of lower velocity. This shearing effect becomes important when the precipitated wax content in the turbulent core is high, that is when the bulk oil temperature has fallen well below the wax appearance point. Burger *et al.* claimed that for neutrally buoyant particles, it is known that particle trajectories near the wall are independent of velocity. Bern *et al.* indicated that an increase in shear rate encourages more wax to disperse towards the pipe wall, and the corresponding increase in wall shear stress may cause the more loosely held deposits to be stripped from the pipe wall.

Burger *et al.* presented a deposition rate expression given in Equation 2 which depends on a parameter called rate deposition constant k^* to quantify the effect of shear dispersion.

$$\frac{dm_s}{dt} = A_i k^* C_w^* \gamma \quad (2)$$

2.3 SURFACE ROUGHNESS

Patton *et al.* (1970) observed that there is no direct correlation between deposition and surface roughness. Jessen *et al.* (1958) concluded that the amount of deposits on a smooth surface is less than that deposited on steel. However, Hunt *et al.* (1962) determined that deposits do not adhere to the metals themselves but are held in place by surface roughness, while Jorda *et al.* (1966) observed that deposition increases with an increase in surface roughness.

3. Experimental Program

An extensive experimental program was initiated to acquire wax deposition data as a function of time, pipe length, flow rate and radial temperature gradient. In this section, detailed descriptions of the experimental test facility as well as the instrumentation installed along the test loop and the data acquisition system are given. A description of the experimental procedures, data acquisition and data interpretation is also included.

3.1 EXPERIMENTAL FACILITY

The experimental facility consists of a flow loop section, a receiver/supply section, and heating and cooling sections.

3.1.1 *Flow Loop*

The flow loop section consists of a 50-m (164-ft) long pipe, having an inner pipe diameter of 43.6 mm (1.71 in), enclosed in an outer concentric pipe of 101.6-mm (4.0-in) diameter. The flow loop is inclined 2° upward over the first 25 m (82 ft) from the inlet, and 2° downward over the last 25 m (82 ft) to enhance drainage of oil and cooling/heating water. The inner test pipe is connected to two horizontal reference pipes, at the inlet and outlet end of the test pipe, each having a length of 3.05 m (10.0 ft) and an inside diameter of 41.1 mm (1.62 in). These reference pipes are heat traced on the outside wall with heating coils through which hot water is circulated. The heating coils are used to heat up the outer pipe wall slightly above the test fluid temperature, to prevent wax deposition in the reference pipes due to molecular diffusion. Two spool pieces or take-out sections are located at a distance of 17.34 m (56.89 ft) and 42.46 m (139.30 ft) from the inlet. A schematic diagram of the 50-m (164-ft) long flow loop section is given in Figure 1.

Oil is circulated in the inner test pipe, and cooling or heating water is circulated counter-current in the concentric pipe. The inlet temperature of the oil is controlled with two shell-tube heat exchanger units. The inlet temperature of the cooling water is controlled by a chilling unit. The maximum working pressure of the test facility is 689 kPa (100 psi).

3.1.2 *Receiver/Supply Section.*

This section consists of a main pump, oil storage tanks, a heater, and a nitrogen purging system. Figure 2 is a schematic diagram of the receiver and supply section. Oil is drained from the storage tanks by a progressing cavity pump with a capacity of 7.5 liters/s (4080 BPD, 120 GPM). A bypass line is used to divert part of the oil flow back to the storage tanks. All oil flow lines in this section are heat traced with heating coils through which hot water is circulated to maintain a wall temperature above the oil temperature. A small heater is used to maintain the temperature of the water. Each of the oil storage tanks has a capacity of 1,000 liters (6.3 bbl). The storage tanks are also maintained at the desired bulk oil temperature with the same heat tracing system. All oil flow lines and the oil storage tanks are insulated.

3.1.3 *Cooling/Heating Sections.*

A schematic diagram of the cooling and heating sections is given in Figure 3. A heating section circulates water that is heated with a large boiler unit and flows through two shell-tube heat exchanger units to control the oil inlet temperature. After each test, hot water is circulated in the concentric pipe to remove or melt all wax accumulated on the walls of the inner test pipe. The maximum temperature attainable is 80 °C (176 °F).

A cooling section circulates water in the outer concentric pipe of the test flow loop. The water is circulated by a recirculation pump and is cooled in a chilling unit. The minimum temperature attainable at the inlet is 4 °C (39.2 °F). A desired temperature is attained by controlling the flow rate of the water flowing through the chilling unit. The cooling and heating water are recovered in separate outlet storage tanks.

3.2 OIL SYSTEM

Crude oil from Mobil Oil Corporation's South Pelto field, Well 10E, in the Gulf of Mexico was used in the experiments. This crude oil has proven wax deposition problems and has been studied extensively by member companies at the DeepStar consortium. The average density of the crude oil was measured by a Micro Motion mass flowmeter. Specific heat capacity (C_{po}) and thermal conductivity (k_o) were determined using Cragoe (1929) expressions for oils. The average C_{po} and k_o, over the range of the test conditions are 1875.67 J/kg-°C (0.4479 Btu/Ibm-°F) and 0.1249 W/m-°C (0.0722 Btu/ft-hr-°F), respectively. Solid weight fraction and solid concentration gradient (dC_w/dT) at various temperatures and pressures were obtained from a Conoco

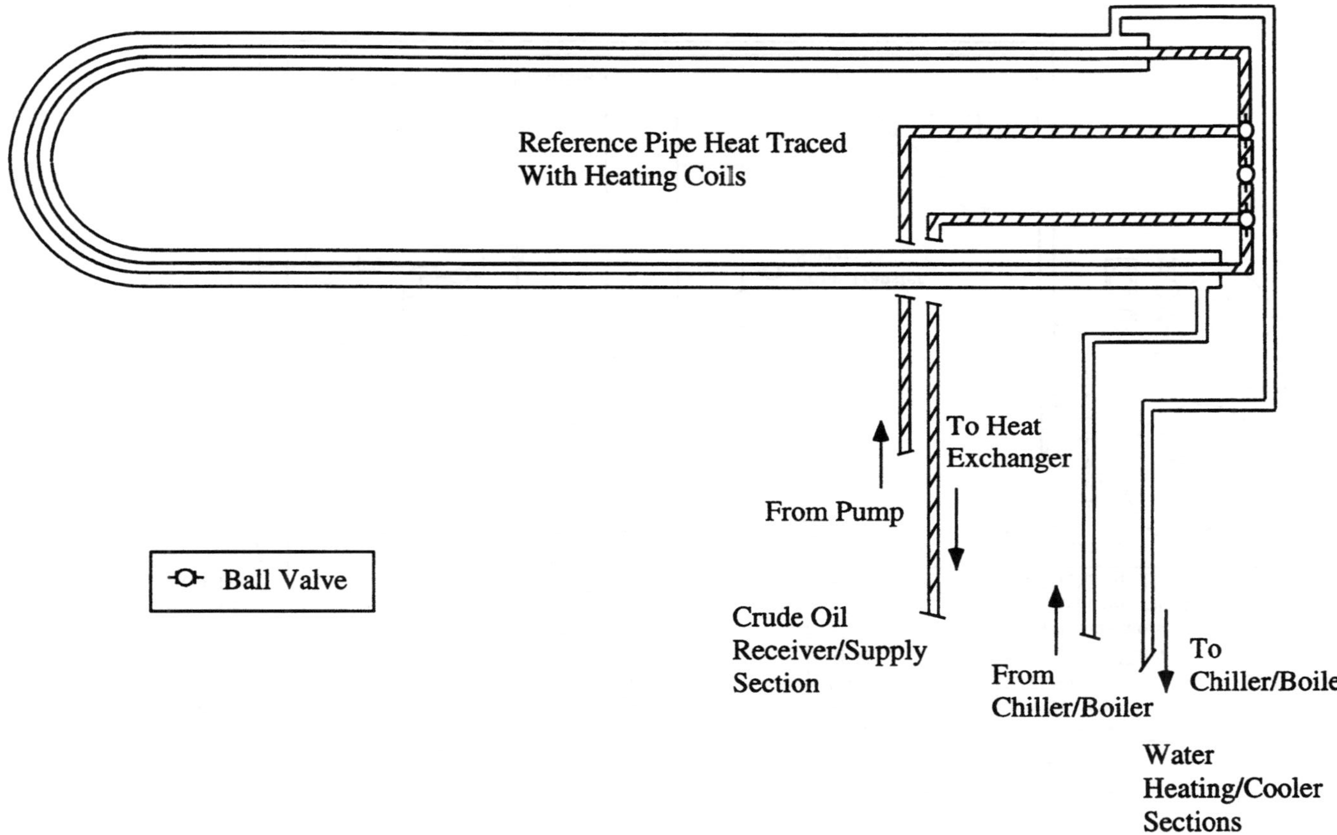

Figure 1. Schematic diagram of the flow loop section

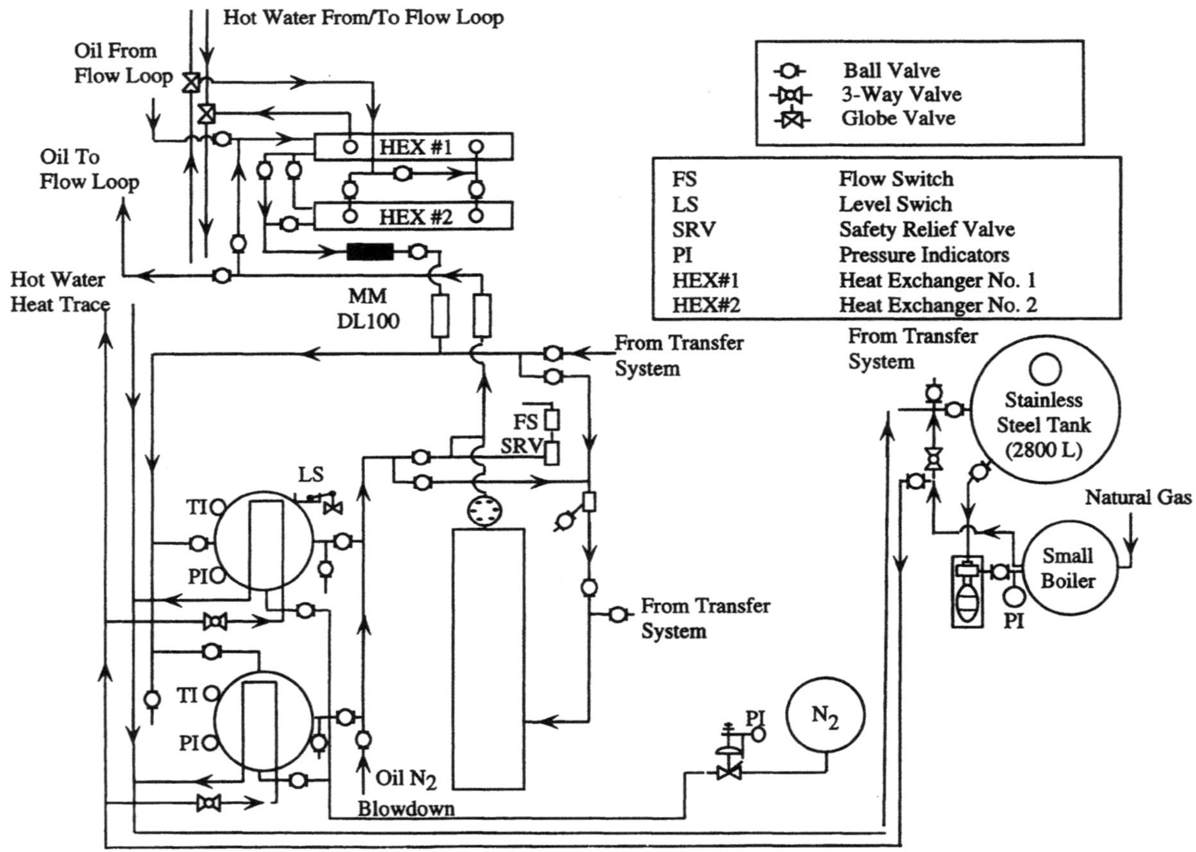

Figure 2. Schematic diagram of the receiver/supply section

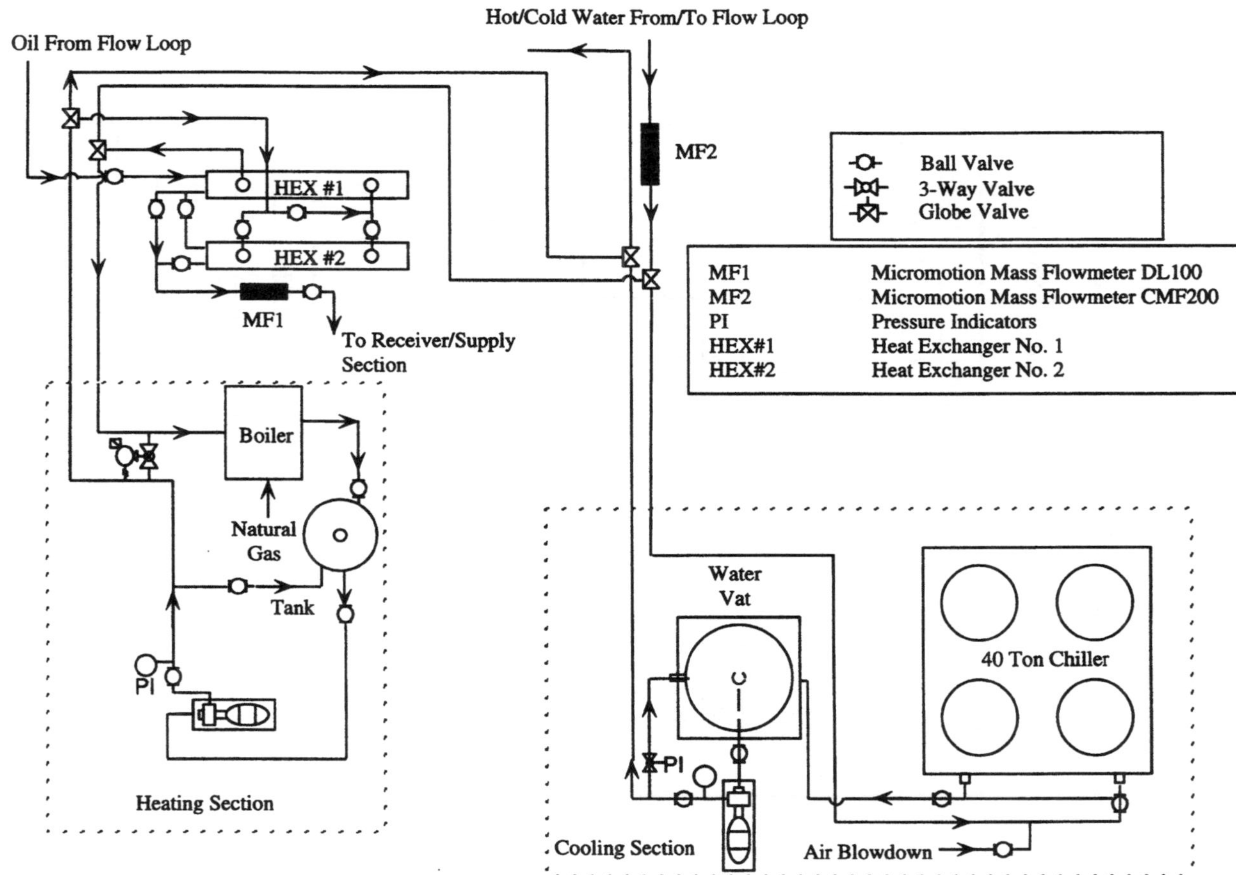

Figure 3. Schematic diagram of the cooling/heating sections

(1996) thermodynamic solid-liquid equilibrium model. Table 1 tabulates these results at various temperatures and pressures.

TABLE 1. Thermodynamic solid-liquid equilibrium of South Pelto crude oil

Temperature ° C	Pressure kPa	Solid Weight %	Solid Gradient dC_w/dT
15.5	103	0.991	9.66E-04
24.0	103	0.547	9.67E-04
32.3	103	0.303	9.94E-04
40.6	103	0.158	1.04E-04
48.9	103	0.063	1.10E-04
15.5	234	0.969	8.50E-04
24.0	234	0.535	8.32E-04
32.3	234	0.297	8.33E-04
40.6	234	0.154	8.47E-04
48.9	234	0.060	8.67E-04

In addition, a Differential Scanning Calorimetry (DSC) analysis was performed by Elf Aquitaine (Volle (1996)) on the same crude oil. Table 2 gives the results of the DSC analysis.

TABLE 2. DSC determination of crystallized fraction by Elf Aquitaine

Temperature, °C	% Weight
45	0.020
40	0.110
35	0.193
30	0.310
25	0.469
20	0.703
15	1.072
10	1.555
5	2.166

3.3 INSTRUMENTATION AND CALIBRATION

The flow rates of crude oil and water were measured with Micro Motion mass flowmeters. Pressures at the inlet and outlet of the inner test pipe, and at the outlet of the pump were measured with absolute pressure transducers. Differential pressures at six (6) locations along the 50-m (164-ft) loop inner test pipe, and at two (2) locations along the reference pipes were measured with differential pressure transducers. Pressures at the inlet and outlet of the outer test pipe were measured with pressure indicators.

Oil temperatures at ten (10) locations along the test pipe were measured using thermocouples. Temperatures of the oil at the inlet and outlet reference pipes, and temperatures of the cooling water at the inlet and outlet of the concentric test pipe were measured using Resistance Temperature Detectors (RTDs). A schematic diagram of the instrumentation is given in Figure 4. In Figure 4, the six (6) locations for the differential pressure measurements along the test pipe are designated by Segment 1 to Segment 6. Table 3 summarizes the instruments used.

All differential pressure transducers were calibrated with a hand held pump connected to a digital manometer. All absolute pressure transducers were calibrated with a portable pressurized tank connected to a mercury manometer. All thermocouples and RTDs were calibrated using a temperature controlled bath.

TABLE 3. Specifications of Instruments

Instrument	Type	Range
DPT1	Foxboro 843DPHOI	0-12.5 kPa
DPT2	Foxboro 843DPHOI	0-12.5 kPa
DPT3	Rosemount 2024 D2A12AOS1B4	0-12.5 kPa
DPT4	Rosemount 2024 D2A12AOS1B4	0-12.5 kPa
DPT5	Rosemount 2024 D2A12AOS1B4	0-12.5 kPa
DPT6	Rosemount 2024 D2A12AOS1B4	0-12.5 kPa
DPT7	Daniel 1151:DP4E12T000B1	0-12.5 kPa
DPT8	Daniel 1151:DP4E12T000B1	0-12.5 kPa
PT1	Foxboro 841GM-BI	689 kPa
PT2	Daniel GP6E12T0007B1	689 kPa
PT3	Daniel GP6E12T0007B1	689 kPa
MF1	Micro Motion DL100	0-21780 kg/hr
MF2	Micro Motion CMF200	0-87100 kg/hr
TT1-TT6	RTD	0-95 °C
TT7-TT16	Thermocouple Type K	0-95 °C

Note: DPTs-Differential Pressure Transducer, PT-Pressure Transducer, MF-Mass Flowmeter, TT-Temperature Transducers

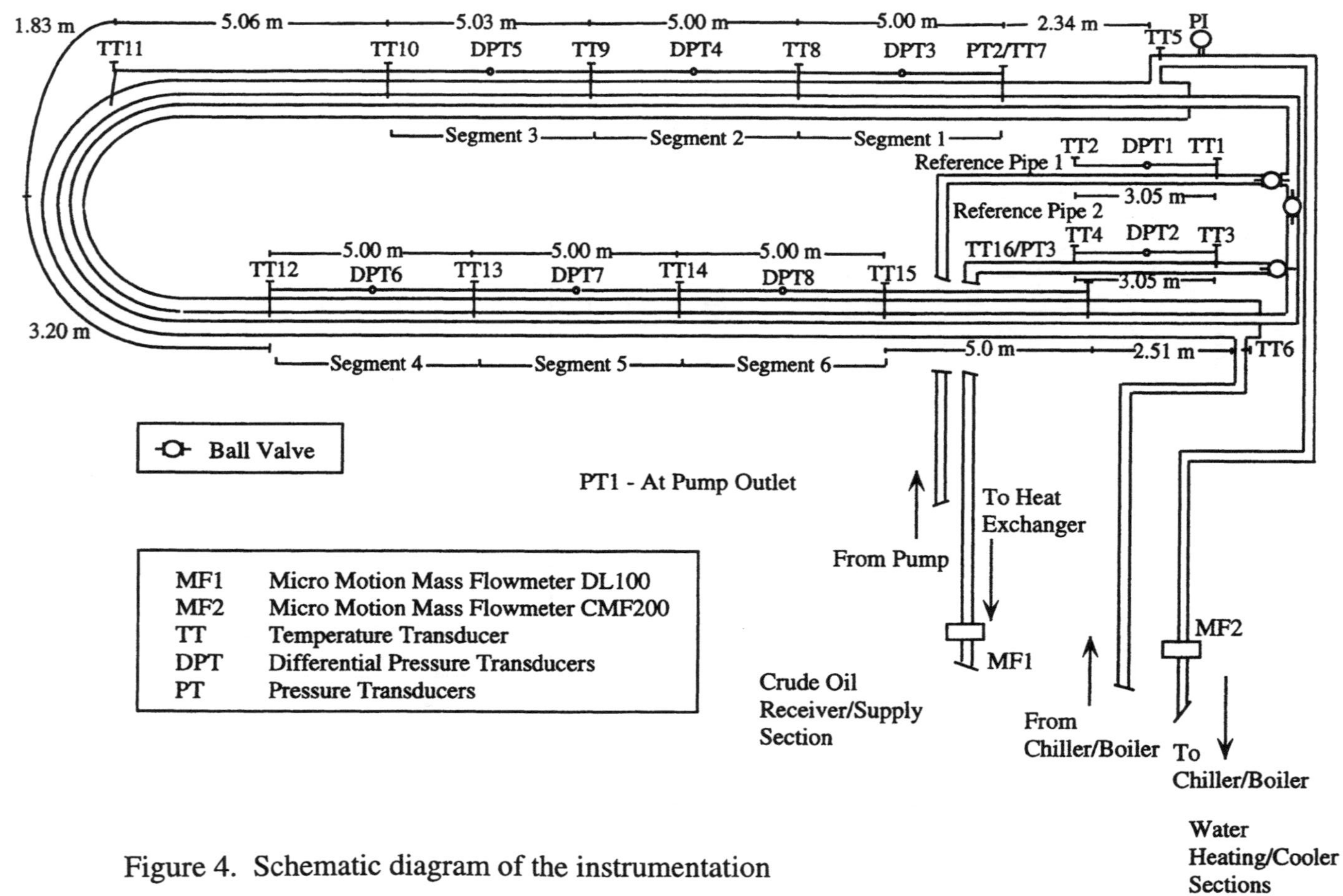

Figure 4. Schematic diagram of the instrumentation

3.4 DATA ACQUISITION/ CONTROL SYSTEM

The hardware for the data acquisition system was composed of a Macintosh computer, an analog multiplexer AMUX-64T, and a high resolution multifunction I/O board model NB-MIO-16H-18. The software was composed of a LabVIEW™ (Version 3.1.1) data acquisition program.

A total of 31 sensor signals, including eight (8) differential pressures, three (3) absolute pressures, sixteen (16) temperature indicators, and two (2) mass flow sensors, each providing measurements of mass flow rate and fluid density, were connected by a Centronics-type wiring to an AMUX-64T analog multiplexer capable of handling up to 32 differential input signals. The multiplexer signals were transferred to a high performance multifunction I/O board, model NB-MIO-16H-18, with an 18 millisecond analog to digital conversion rate and capable of acquisition sampling rates up to 55,000 samples/s. All instrument signals were in a format of either -10 to +10 volts. Continuous storage of the acquired data was handled by the LabVIEW™ software interface. LabVIEW™ employs a graphical programming language that allows users to create on-screen virtual instruments that can monitor the variation of each output signal. LabVIEW™ converted the output signals into real time pressures, temperatures and flow rates at specified time intervals. These data were stored in an ASCII formatted file.

3.5 EXPERIMENTAL TESTS

A total of four tests were conducted as shown in Table 4. The tests covered a wide range of operating conditions with inlet oil temperatures ranging from 37.8 °C (100.0 °F), which was below the cloud point temperature to 54.5 °C (130.0 °F), which was above the cloud point temperature. The cloud point temperature was determined experimentally by DiEspoti and Tackett (1996) to be 48.9 °C (120.0 °F). The oil flow rate ranged from 2.76E-4 m^3/s (150 BPD) (laminar flow) to 2.76E-3 m^3/s (1500 BPD) (turbulent flow). The inlet water temperature was 15.6 °C (60 °F) in all cases.

TABLE 4: Experimental cases

	Test 1	Test 2	Test 3	Test 4
Oil Flow (m^3/s)	2.76E-3	9.2E-4	2.76E-4	9.2E-4
Oil Velocity (m/s)	1.853	0.6187	0.1853	0.6187
Oil Inlet Temp (°C)	54.5	54.5	54.5	37.8
Oil Inlet Reynolds No.	12403	4134	1240	3064
Water Inlet Temp (°C)	15.6	15.6	15.6	15.6
Water Velocity (m/s)	0.33	0.33	0.33	0.33
Water Inlet Reynolds No.	16440	16440	16440	16440
Test Run Time (hrs)	22	18	44	13

3.6 EXPERIMENTAL PROCEDURE

The first task was to conduct a check of all instruments using the data acquisition system. Initially, the South Pelto crude oil was bypassed around the test loop and flowed only through the heat exchangers. The desired inlet temperature of the oil was achieved by controlling the flow rate of the hot water flowing into the heat exchangers. At the same time, the water at an inlet temperature of 15.6 °C (60.0 °F) was circulated through the outer pipe of the test section.

Once the desired inlet oil temperature was achieved, a bypass valve was closed to allow the oil to flow into the test loop. Continuous data acquisition was initiated, and real time measurements were recorded. The oil inlet temperature was maintained by adjusting the flow rate of the heating water flowing into the heat exchangers. The cooling water inlet temperature was maintained by adjusting the flow rate of the water flowing into the chiller.

3.7 WAX DEPOSITION DATA INTERPRETATION

3.7.1 *Reference Pipes*

An indirect method using temperature and pressure drop measurements at the inlet and outlet reference pipes, and at various locations along the 50-m (were-ft) long test loop can be used to calculate the axial wax deposition thickness.

The procedure described assumes that the crude oil behaves as a Newtonian fluid. This assumption is based on the fact that the relatively small amount of wax out of solution over the temperature range of the present study did not have a significant impact on the behavior of the crude oil. Results of the differential scanning calorimetry analysis given in Table 2 show only 0.3% of wax out of solution at 30.0 °C (86.0 °F). Measurements made in the reference pipes were used to verify this assumption. Assuming Newtonian behavior, the procedure for estimating paraffin deposition thickness is described below.

From a pressure drop measurement, the Fanning (Welty *et al.* (1983)) friction factor in each reference pipe (no wax deposition) can be expressed as follows:

$$f = \left.\frac{dp}{dx}\right|_{fR} \left(\frac{d_R}{\rho_o v^2} \right) \frac{1}{2} \tag{3}$$

Using the above friction factor, a Reynolds number, N_{Re}, is obtained from appropriate Newtonian fluid friction factor expressions. For laminar flow,

$$f = \frac{16}{N_{Re}} \tag{4}$$

The following explicit correlation by Zigrang and Sylvester (1985) is used to obtain N_{Re} for turbulent flow:

$$\frac{1}{\sqrt{4f}} = -2\log\left\{\frac{2\varepsilon/d_R}{3.7} - \frac{5.02}{N_{\text{Re}}}\log\left(\frac{2\varepsilon/d_R}{3.7} + \frac{13}{N_{\text{Re}}}\right)\right\} \tag{5}$$

Assuming crude oil density is constant along the reference pipe, the apparent in-situ viscosity can be determined from the definition of N_{Re}. Since temperature was also measured, the crude oil viscosity as a function of temperature can be obtained.

3.7.2 *Test Section Pipe*

The effective diameter, d_w, as a result of wax buildup in a segment of the test pipe through which oil is flowing at a volumetric $\tilde{Q}$ is given by:

$$d_w = \left[\frac{2f\rho_o\left(\frac{\tilde{Q}}{\pi/4}\right)^2}{\left.\frac{dp}{dx}\right|_{fT}}\right]^{1/5} \tag{6}$$

For simplicity, f can be calculated from a smooth pipe expression given by the Blasius formula:

$$f = BN_{\text{Re}}^{-m}\begin{Bmatrix}\text{B} = 16, \text{m} = 1; \text{laminar} \\ \text{B} = 0.046, \text{m} = 0.2; \text{turbulent}\end{Bmatrix} \tag{7}$$

The smooth pipe assumption is supported by the following:

- The extent of the effective roughness as wax builds up on the pipe wall is unknown. Currently, there are no methods available for predicting this effective roughness.
- As wax builds up on the pipe wall, the roughness of the pipe depends upon the nature of the wax deposited on the pipe wall. The surface of the wax layers is assumed to be uniform and is treated as smooth. Single phase non-depositional tests indicated that a relative roughness for the pipe was approximately 0.0009. For the range of Reynolds numbers encountered (1,240 to 12,403), there is little difference between rough pipe and smooth pipe friction factors. This is true even if a somewhat larger relative roughness is used. Thus, the extent of the effective roughness due to paraffin deposition can be considered to be negligible at the current test conditions.

Substituting Equation 4 for laminar flow into Equation 6 yields Equation 8.

$$d_w = d_i - 2\delta = \left[\frac{\frac{8\pi\mu_o}{\tilde{Q}}\left(\frac{\tilde{Q}}{\pi/4}\right)^2}{\left.\frac{dp}{dx}\right|_{fT}}\right]^{1/4} \tag{8}$$

Substituting the Blasius friction factor expression for turbulent flow, Equation 7, into Equation 6 yields Equation 9.

$$d_w = d_i - 2\delta = \left[\frac{2(0.046)\left(\frac{\pi\mu_o}{4\rho_o\tilde{Q}}\right)^{0.2}\rho_o\left(\frac{\tilde{Q}}{\pi/4}\right)^2}{\left.\frac{dp}{dx}\right|_{fT}}\right]^{1/4.8} \tag{9}$$

All parameters inside the brackets of Equations 8 and 9 were recorded at each time interval. This allowed determination of the wax thickness at each time interval. Prior to the wax thickness calculation, differential pressure measurements at each test pipe segments were corrected for the pressure loss due to elevation to obtain the pressure loss due to friction, $\left.\frac{dp}{dx}\right|_{fT}$.

4. Experimental Results And Model Comparisons

In this section, experimental results are presented and discussed for only Test No. 1. Due to the proprietary nature of the project, analyses of other test results cannot be given. Measured temperature profiles and calculated wax buildup history as a function of pipe length are presented.

4.1 TEMPERATURE PROFILES

Bulk oil and water temperatures at initial (average of first 30 minutes) and final (average of last 30 minutes) times were plotted for Test No. 1 and are shown in Figure 5. It can be observed that:

- The initial bulk oil temperature drops from 54.5 °C (130.0 °F) at inlet to about 43.0 °C (109.4 °F) at outlet.
- The bulk oil outlet temperatures at final conditions had increased to a higher value than at initial conditions.

- The cooling water temperature profile at final conditions is slightly lower than at initial conditions.
- The measured temperature profiles suggest that an insulation effect is taking place as wax is being deposited along the pipe wall.

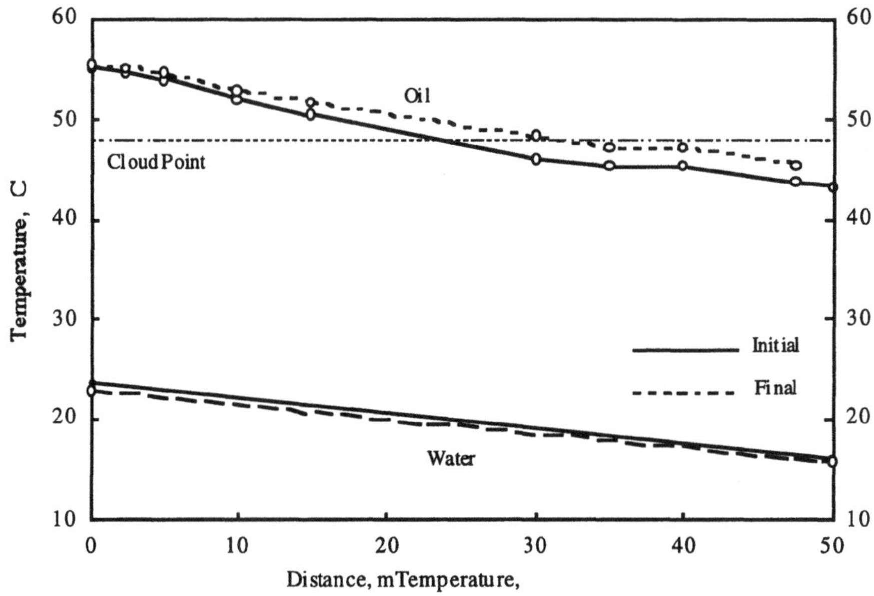

Figure 5. Temperature profiles at initial and final times for Test 1 { $\tilde{Q}$ = 2.76E-3 m^3/s (1500 BPD), $T_{o,in}$ = 54.5 °C (130.0 °F), $T_{cw,in}$ = 15.6 °C (60.0 °F)}

4.2 WAX BUILDUP HISTORY

Wax buildups at 5-min intervals were determined along the 5-m (16.4-ft) long differential pressure measurement segments (Segments 1, 2, 3, 4, 5, and 6) by an iterative algorithm based on the definition of frictional pressure drop described in the wax deposition data interpretation section. These differential pressure measurement segments are located at a distance of 2.34 m (7.68 ft) to 7.34 m (24.08 ft) (Segment 1), 7.34 m (24.08 ft) to 12.34 m (40.49 ft) (Segment 2), 12.34 m (40.49 ft) to 17.34 m (56.89 ft) (Segment 3), 27.46 m (90.09 ft) to 32.46 m (106.49 ft) (Segment 4), 32.46 m (106.49 ft) to 37.46 m (122.90 ft) (Segment 5), and 37.46 m (122.90 ft) to 42.46 m (139.30 ft) (Segment 6) from the inlet of the inner pipe. Wax buildup values vs. time were plotted for each pipe segment and are shown in Figs. 11-16. Apparent fully developed thickness or average wax thickness at final time and initial wax buildup rates are noted.

Test 1 lasted 22 hrs. It can be observed in Figs. 6-11 that wax thickness increases steadily, reaching an apparent fully developed thickness after some period of time. The apparent fully developed wax thickness or wax thickness at final time for Segments 1 (Figure 6), 2 (Figure 7), 3 (Figure 8), 4 (Figure 9), and 6 (Figure 10) are 0.6 mm (0.024 in), 0.5 mm (0.019 in), 0.38 mm (0.015 in), 0.5 mm (0.019 in), and 0.6 mm (0.024 in), respectively.

The times to achieve the apparent fully developed wax thickness vary among the segments. The initial wax buildup rate, which can be obtained from the initial slope of the plots, also varies in all segments. Comparatively, the times to achieve apparent fully developed wax thickness in Segments 1, 2 and 3 are longer than in Segments 4 and 6. The initial wax buildup rates for Segments 1, 2, 3, 4 and 6 are 0.45 mm/hr (0.018 in/hr), 0.45 mm/hr (0.018 in/hr), 0.19 mm/hr (0.007 in/hr), 0.11 mm/hr (0.004 in/hr) and 0.11 mm/hr (0.004 in/hr), respectively. Segments 1, 2 and 3 give higher initial wax buildup rates as compared to Segments 4 and 6. This is due to lower radial temperature gradients and possibly wax depletion in the oil as time progresses.

The wax thickness in Segments 4 and 6 decreased suddenly after a period of 17 hrs. The sudden drop in wax thickness could possibly be due to sloughing of wax from the pipe wall. It should be mentioned that, based on the temperature profiles presented in Figure 5, the average temperatures in Segments 1, 2 and 3 are above the cloud point temperature, and the average temperatures in Segments 4, and 6 are at or below the cloud point temperature.

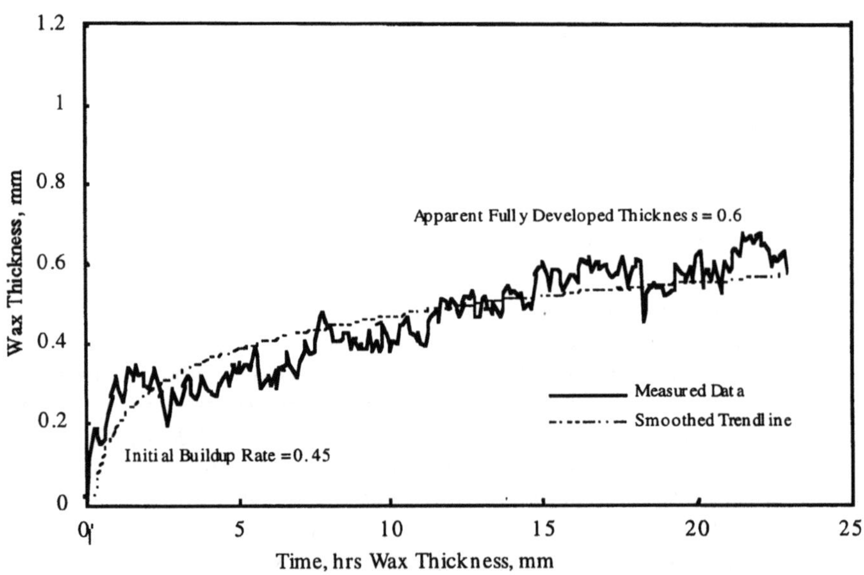

Figure 6. Wax buildup profile for Segment 1 {2.34 m to 7.34 m (7.68 ft to 24.08 ft)} - Test 1

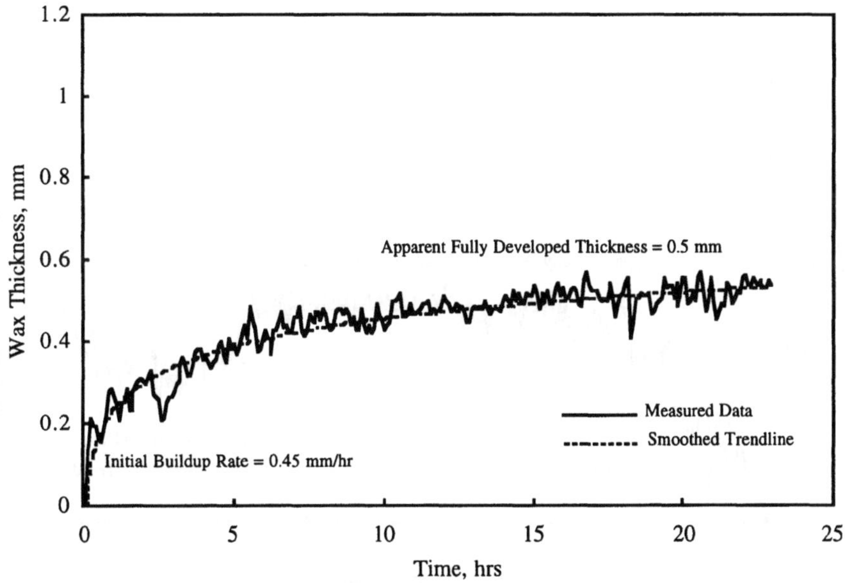

Figure 7. Wax buildup profile for Segment 2 {7.34 m to 12.34 m (24.08 ft to 40.49 ft)} - Test 1

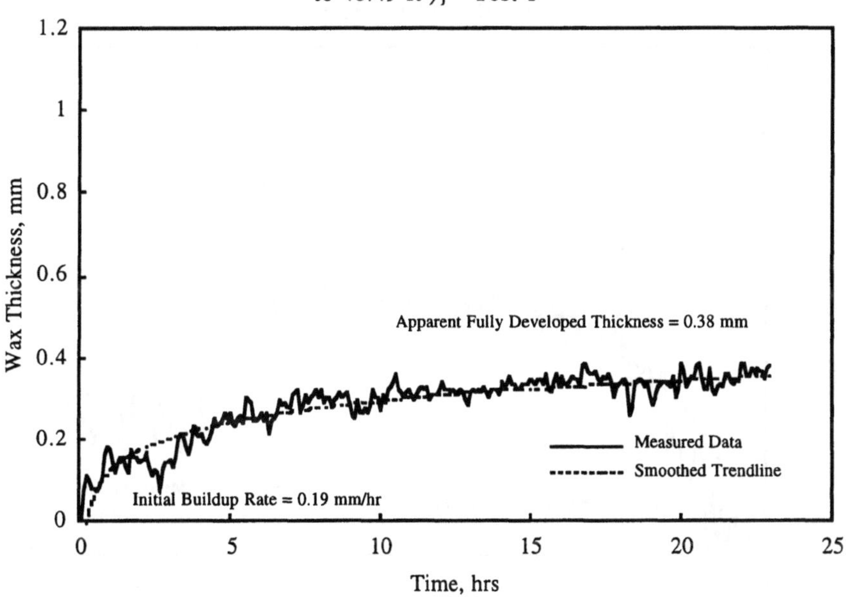

Figure 8. Wax buildup profile for Segment 3 {12.34 m to 17.34 m (40.49 ft to 56.89 ft)} - Test 1

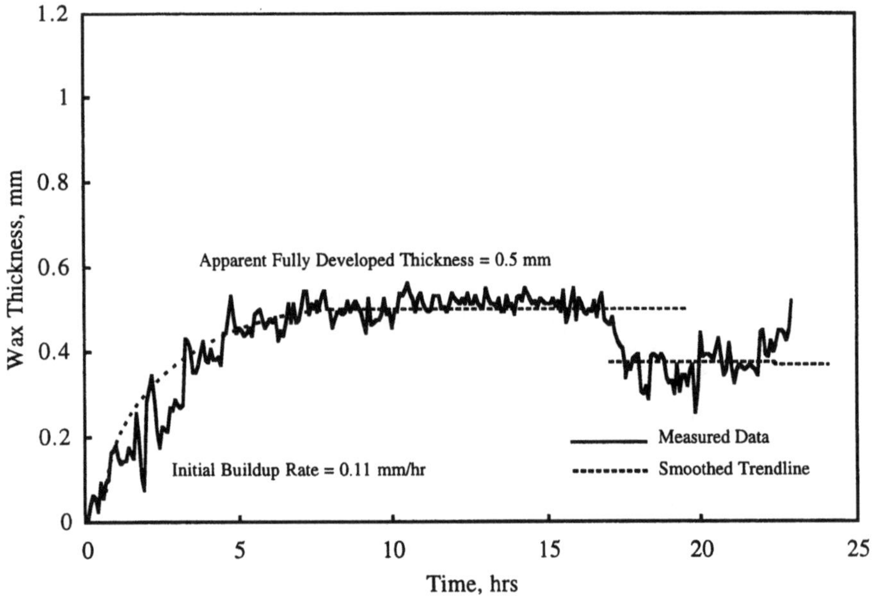

Figure 9. Wax buildup profile for Segment 4 {27.46 m to 32.46 m (90.09 ft to 106.49 ft)} - Test 1

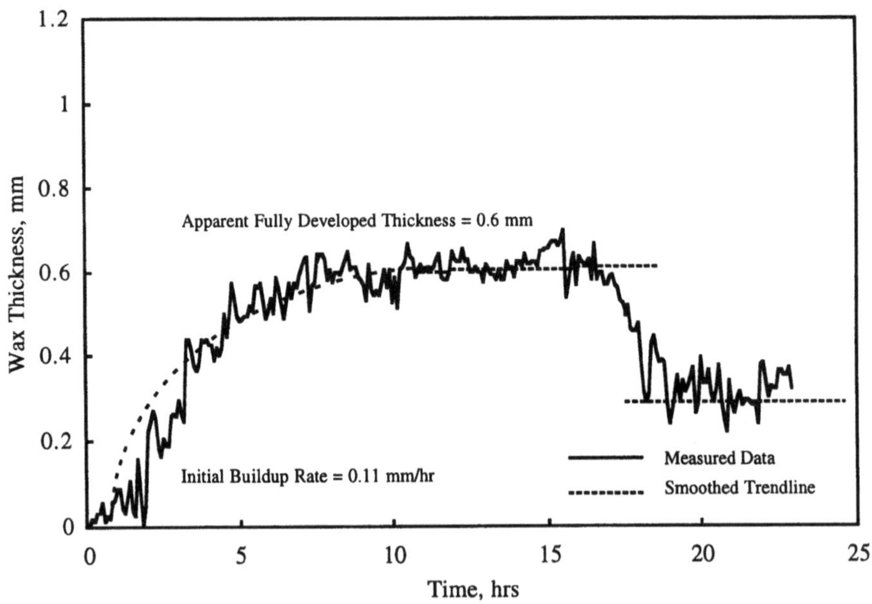

Figure 10. Wax buildup profile for Segment 6 {37.46 m to 42.46 m (122.90 ft to 139.30 ft)} -Test 1

During the test period, it was observed that the differential pressure instrument at Segment 5 was malfunctioning. The malfunctioning could be due to plugging of pressure line ports. As shown in Figure 11, the buildup trend in this segment was not well defined.

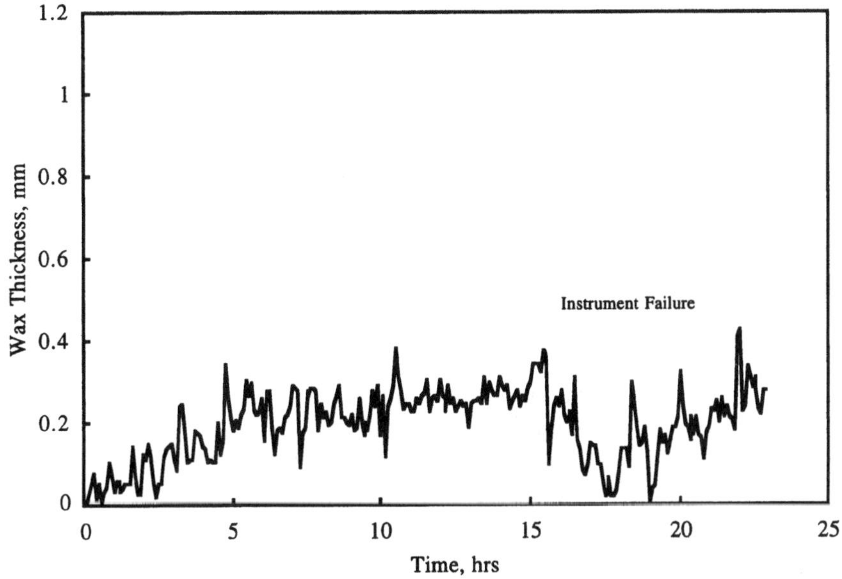

Figure 11. Wax buildup profile for Segment 5 {32.46 m to 37.46 m (106.49 ft to 122.90 ft)} -Test 1

5. Observations

Based on the preliminary experimental program and analysis of the preliminary data generated in this first stage of single phase paraffin deposition investigation, the following observations were noted:

- A new single phase paraffin deposition test facility was assembled and operated. The test facility was comprised of four main components: flow loop, receiver and supply, cooling and heating sections. The test facility was proven to be capable of depositing wax on the inner pipe wall of the test pipe.
- An insulation effect was observed as wax was being deposited along the pipe wall. The amount of wax deposited could be correlated from the extent of the wax insulation effect.
- Wax buildup increased steadily and reached an apparent fully developed thickness after some period of time.

- Initial wax buildup rate was higher at the inlet than at the outlet.
- The differential pressure technique with appropriate assumptions can be used with confidence to estimate the wax thickness for a single phase flow system.

In addition, at the conclusion of Test No. 4, the oil and water were drained from the test section, a take-out section was removed, and the deposited wax was visually inspected. The wax observed had a thickness similar to what was predicted from pressure drop measurements. The deposited wax had the appearance of a viscous oil, and subsequent analysis confirmed that approximately 80 to 90% of the deposition was trapped oil.

6. Nomenclature

Symbol	Description	
A_i	=	inner surface area of pipe wall, (m^2)
B	=	friction factor coefficient
C_w	=	concentration of wax in solution, (weight %)
C_w^*	=	concentration of wax at pipe wall, (weight %)
D_m	=	molecular diffusion coefficient, (m^2/s)
d_i	=	inside pipe diameter, (m)
d_w	=	inside pipe diameter as a result of wax buildup, (m)
d_R	=	reference pipe diameter, (m)
f	=	friction factor
k*	=	shear dispersion deposition rate constant, (kg/m^2)
m_m	=	deposition by diffusion, (kg)
m_s	=	deposition by shear, (kg)
m	=	friction factor coefficient
N_{Re}	=	Reynolds number
p	=	pressure, (kPa)
$\tilde{Q}$	=	volumetric flow rate, (m^3/s)
r	=	radial distance, (m)
T	=	temperature, (°C)
t	=	time, (s)
v	=	velocity, (m/s)
$\left.\frac{dp}{dx}\right\rvert_{fR}$	=	frictional pressure drop in reference pipe, (Pa/m)
$\left.\frac{dp}{dx}\right\rvert_{fT}$	=	frictional pressure drop in test pipe, (Pa/m)

Greek letters		Description
ρ_w	=	density of wax, (kg/m^3)

ρ_o	=	oil density, (kg/m^3)
μ_o	=	oil viscosity, (kg/m s)
γ	=	shear rate, (1/s)
δ	=	wax thickness, (m)
ε	=	roughness, (m)

7. References

1. Bern, P. A., Winthers, W. R. & Cairns, J.R. : "Wax Deposition in Crude Oil Pipelines," European Offshore Petroleum Conference & Exhibition, London, UK, 21-24, (1980).
2. Burger, E. D., Perkins, T.K. & Striegler, J.H.: "Studies of Wax Deposition in the Trans- Alaska Pipeline," J. of Petroleum Technology, 1075-1086, (June 1981).
3. Cole, R. J, and Jessen, F.W.: "Paraffin Deposition," Oil and Gas Journal, vol 58, pp 87, (Sept 1960).
4. Cragoe, N.B.S. Misc. Publ., pp 97, (1929) cited in Perry & Chilton: "Chemical Engineers Handbook", pp. 6-237 & pp 6-243 (1973).
5. DiEspoti, E.J. and Tackett, J.E. : "Fluid Characterization and Property Evaluation for Paraffin Deposition in Multiphase Flowlines and Wellbores," Quarterly progress report, (September 23, 1996).
6. Hunt Jr., B. E.: "Laboratory Study of Paraffin Deposition," JPT, 1259-1269, (November 1962).
7. Jessen, F.W. & Howell, J.N. : "Effect of Flow Rate on Paraffin Accumulation in Plastic, Steel, and Coated Pipe," Petroleum Transactions, AIME, Vol. 213, 80-84, (1958).
8. Jorda, R. M. : "Paraffin Deposition and Prevention in Oil Wells," JPT, 1605-1612, (December 1966).
9. Matzain, A., "Single Phase Liquid Paraffin Deposition Modeling," MS Thesis, The U. of Tulsa (1997).
10. Patton, C.C., & Casad, B.M. : "Paraffin Deposition from Refined Wax-Solvent System," SPEJ, pp 17, (March 1990).
11. Volle, J.P. : "DSC Determination of Crystallized Fraction of South Pelto Crude Oil by Elf Aquitaine", (1996).
12. Welty, J. R., Wicks C. E., Wilson, R.E.: "Fundamentals of Momentum, Heat and Mass Transfer," John Wiley and Sons, 3rd Edition, pp 202-211, (1983).
13. Zigrang, D. J. and Sylvester, N.D.: " A Review of Explicit Friction Factor Equations," Trans. ASME, JERT, vol 107, pp 280-283, (1985).
14. Conoco PARAFFIN Prediction Program and Manual.

LEVITATION OF CORE FLOWS

C. MATA, R. BAI & D. D. JOSEPH
University of Minnesota
Department of Aerospace Engineering and Mechanics
107 Akerman Hall, 110 Union Street
Minneapolis, MN 55455, USA

Abstract

A simple model is proposed for a 2D horizontal core annular flow in which the effect of gravity due to the difference in the densities of the two fluids is the eccentricity of the core. We split the domain through the center of the core; we characterized each sub domain by means of local variables, for instance the holdup ratio. We found that the smallest global pressure drop is achieved if and only if the local holdup ratios are equal to the global holdup ratio (for a perfect core annular flow this only happens when the core is centered). We used this result in a direct simulation of spatially periodic 2D wavy core annular flows carried out under the assumption that the viscosity of the oil core is so large that it moves as a rigid solid which may nevertheless be deformed by pressure forces in the water. The waves that develop are asymmetric with steep slopes in the high pressure region at the front face of the wave crest and shallower slopes at the low pressure region at lee side of the crest, as Bai et al. describe (1996). However two new issues are confronted in our 2D simulation. First, the shape and length of the upper and lower waves are different. Second, the displaced 2D core can be thought to represent eccentricity. We conclude that a positive pressure force is required to levitate the core off the wall when the densities are not matched and that the difference in the upper and lower wave shapes restore the effect of the eccentricity by allowing the local holdup ratios to be equal to the global holdup ratio; thus the pressure drop is the smallest possible.

---

1. Introduction

Lubricated pipelining of viscous materials like heavy crude, slurries and capsules is robustly stable and has a high economic potential. The viscous material does not touch the wall. After Joseph and Renardy (1993) the drag reduction in the case of crude oil, which can be achieved by lubrication is of the order of the viscosity ratio with increased throughputs of ten thousand or more. These lubricated flows are called core flows because the viscous material flows in a core lubricated all around by water.

A surprising property of core flows is that the flow in a horizontal line will lubricate with the core lubricated off the wall even if the core is lighter or heavier than the lubricating water. This levitation could not take place without a hydrodynamic lifting action due to waves sculpted on the core surface. In the case of very viscous liquids, the waves are basically standing waves which are convected with the core as it moves downstream. This picture suggests a lubrication mechanism for the levitation of the core analogous to mechanisms which levitate loaded sliders leavings at low Reynolds numbers. Ooms et al.

[1984] and Oliemans and Ooms (1986) gave a semi empirical model of this type and showed that it generated buoyant forces proportional to the first power of the velocity to balance gravity.

Feng, Huang and Joseph (1995) suggested that the conventional mechanism of lubrication can not describe the water lubricated pipelining of crude oil. They predicted that the high pressures on the forward facing slope of the wave, where the water enters into a converging gap, steepens the wave and the low pressure on the lee side smoothes the wave there. This leads to unsymmetric waves which are actually nearly opposite to the ones which are necessary to levitate a slipper bearing. They also predicted the existence a threshold Reynolds number corresponding to a change in the sign of the pressure force on the core, from suction at Reynolds number below the threshold, as in the reversed slipper bearing in which the slipper is sucked to the wall, to compression for Reynolds number greater than the threshold as in flying core flow in which the core can be pushed off the wall by stagnation pressure.

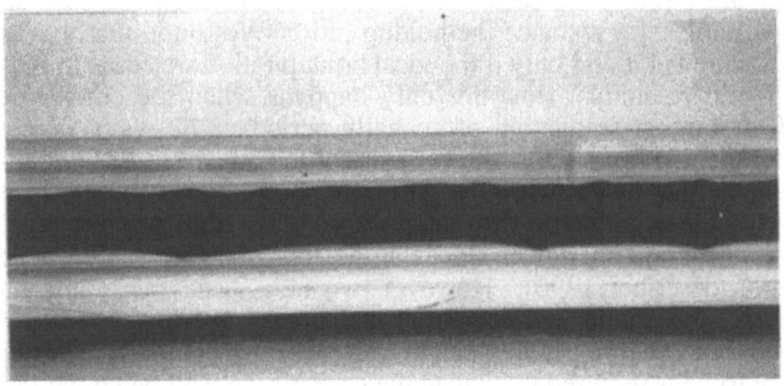

Figure 1.1 Wavy core annular flow in a 1" diameter glass pipe for Reynolds number $\Re$=533, oil to water volume ratio η=0.9 and holdup ratio h=1.4. The upper and lower wave lengths are clearly different.

Figure 1.1 presents a wavy core annular flow in a 1" diameter glass pipe. The wave shape definitely looks like the flying core flow described by Feng et al. (1995). However, the upper and lower wave lengths are clearly different. This difference is due to the buoyancy force acting on the core, because the oil and water densities are not matched. Moreover, the overall effect is an eccentric core flow, which is a 3D problem in a pipe.

Bai, Kelkar and Joseph (1996) performed a direct simulation of interfacial waves in a high viscosity ratio and axisymmetric core annular flow, in which the water and oil densities were matched. In this case the problem of levitation does not arise. However, they computed the pressure distribution that actually develops on the wave and they found that it is such as to levitate a slightly displaced core only when the Reynolds number is greater than a threshold value which depends on the involved parameters. Lubrication by inertia was suggested from this analysis.

The purpose of this work was to carry out a direct simulation of core flows under gravity when the densities are not matched. For simplicity a 2D domain was chosen.

2. 2D perfect core flow

What we call 2D perfect core flow is actually an ideal situation in which two fluids (fluid 1 and fluid 2) with the same density and different viscosities ($\frac{\mu_2}{\mu_1} < 10^{-5}$) flow down an infinite two dimensional horizontal pipeline in a core annular configuration (see figure 2.1), where the interfaces are perfectly smooth and the core is not necessarily centered. We assume that there is a connection somewhere between the upper and lower annuli.

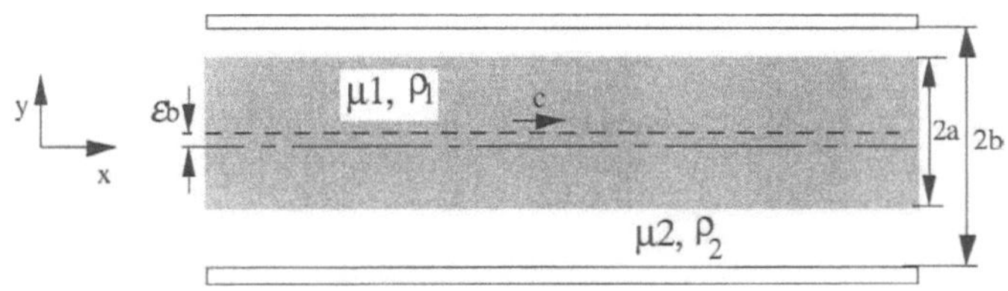

Figure 2.1 Eccentric 2D perfect core flow ($\rho_1=\rho_2$ and $\frac{\mu_2}{\mu_1} < 10^{-5}$). Here ε is a dimensionless measure of the eccentricity.

The pressure field is given by equation (2.1) in the core and, by equation (2.2) in the upper and lower annuli.

$$P_1 = \gamma x - \rho_1 g y + C_1, \tag{2.1}$$

$$P_2 = \gamma x - \rho_2 g y + p(y) + C_2. \tag{2.2}$$

Notice that $\frac{dP_1}{dx} = \frac{dP_2}{dx} = \frac{dP}{dx} = \gamma$ is a unique pressure gradient within the whole domain.

The velocity field in the regions $-b \le y \le -a$ *and* $a \le y \le b$ can be obtained from

$$\frac{\partial u}{\partial x} = 0$$
$$\frac{\partial^2 u}{\partial y^2} = 2\Gamma \tag{2.3}$$

where $\Gamma = -\frac{\gamma}{2\mu_2}$. Since both fluids have the same density, no buoyancy force is present and the position of the core is not fixed.

2.1 SYMMETRIC 2D PERFECT CORE FLOW (PCF)

The solution of equations (2.3) for the symmetric 2D PCAF will be now presented as a function of the geometric parameters shown in figure 2.2, for which the eccentricity ε is zero, and the pressure gradient $\frac{dP}{dx} = \gamma$. The boundary conditions for the velocity are $u = c,\ \ v = 0$ in the region $-a \le y \le a$, And $u = c \ \ v = 0$ at $y = -b \ \ y = b$. The shear stress condition at $y = -a$ and $y = a$ is $\mu_1 \frac{\partial u}{\partial y} = \mu_2 \frac{\partial u}{\partial y}$. Obviously, terms involving the ratio $\frac{\mu_2}{\mu_1} < 10^{-5}$ were neglected.

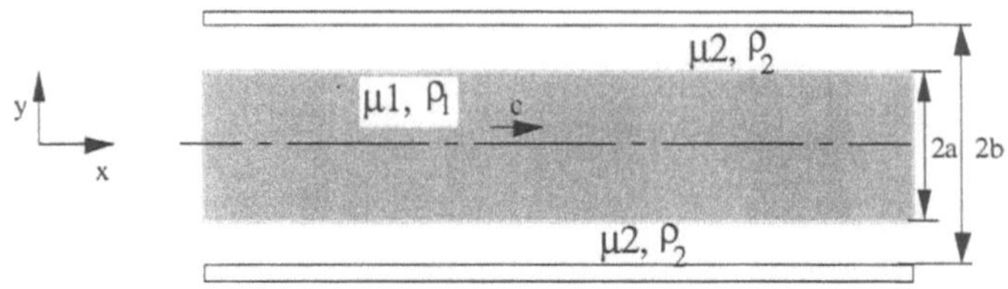

Figure 2.2 Symmetric 2D perfect core flow ($\rho_1 = \rho_2$ and $\frac{\mu_2}{\mu_1} < 10^{-5}$).

The core velocity *c* is:

$$c = \Gamma b^2 \left(1 - \eta^2\right) \tag{2.4}$$

where $\eta = \frac{a}{b}$ and $\Gamma = -\frac{\gamma}{2\mu_2}$.

The total flow rate of fluid 1 is:

$$Q_1 = 2ac = 2\Gamma b^3 \eta \left(1 - \eta^2\right) \tag{2.5}$$

The total flow rate Q_2 and average velocity V_2 for fluid 2 are given by equations (2.6) and (2.7), respectively.

$$Q_2 = \tfrac{2}{3} \Gamma b^3 \left(2 - 3\eta + \eta^3\right) \tag{2.6}$$

$$V_2 = \frac{Q_2}{2b(1 - \eta)} \tag{2.7}$$

The holdup ratio *h*, is convenient parameter defined as the ratio of the average velocities of fluid 1 and fluid 2. It can be expressed as follows:

$$h = \frac{c}{V_2} = \frac{3(1-\eta)\left(1-\eta^2\right)}{2 - 3\eta + \eta^3} \tag{2.8}$$

Notice that $\lim_{\eta \to 0} h = 1.5$ and $\lim_{\eta \to 1} h = 2$. These limits correspond to an ideal core flow. However, this result allows us to set the upper limit of the holdup ratio in a real situation, and this would be between 1.5 and 2.

2.2 ECCENTRIC 2D PERFECT CORE FLOW (PCF)

The solution of equations (2.3), with β instead of γ as the pressure gradient or $\frac{dP}{dx}=\beta$, will be now presented as a function of the geometric parameters shown in figure 2.1, on which the frame of reference is located in the center of the core. The boundary conditions for the velocity are $u=c \quad v=0$ in the region $-a\le y\le a$, And $u=c \quad v=0$ at $y=-b \quad y=b$. The shear stress condition at $y=-a$ and $y=a$ is $\mu_1\frac{\partial u}{\partial y}=\mu_2\frac{\partial u}{\partial y}$.

Again, terms involving the ratio $\frac{\mu_2}{\mu_1}<10^{-5}$ were neglected.

In this case, the core velocity is:

$$c=Bb^2\frac{(1+\eta)}{(1-\eta)}\left\{(1-\eta)^2-\varepsilon^2\right\} \tag{2.9}$$

where ε is a dimensionless measure of the eccentricity of the core and $B=-\frac{\beta}{2\mu_2}$. Notice that $\lim_{\varepsilon\to 0} c=Bb^2\left(1-\eta^2\right)$.

Let

$$K=\frac{c}{Bb^2}=\frac{(1+\eta)}{(1-\eta)}\left\{(1-\eta)^2-\varepsilon^2\right\} \tag{2.10}$$

The total flow rate of fluid 1 is:

$$Q_1=2ac=2Bb^3\eta K \tag{2.11}$$

For the upper half, the fluid 2 flow rate Q_{2u} and average velocity V_{2u} are given by equations (2.12) and (2.13), respectively

$$Q_{2u}=\tfrac{1}{2}Bb^3\{(1-\eta)-\varepsilon\}\left\{\tfrac{1}{3}\left[(1-\eta)\{(1-\eta)-2\varepsilon\}+\varepsilon^2\right]+K\right\} \tag{2.12}$$

$$V_{2u}=\frac{Q_{2u}}{b(1-\eta)} \tag{2.13}$$

For the lower half, the fluid 2 flow rate Q_{2l} and average velocity V_{2l} are given by equations (2.14) and (2.15).

$$Q_{2l}=\tfrac{1}{2}Bb^3\{(1-\eta)+\varepsilon\}\left\{\tfrac{1}{3}\left[(1-\eta)\{(1-\eta)+2\varepsilon\}+\varepsilon^2\right]+K\right\} \tag{2.14}$$

$$V_{2l}=\frac{Q_{2l}}{b(1-\eta)} \tag{2.15}$$

Notice that $\lim_{\varepsilon\to 0} Q_{2u}=\lim_{\varepsilon\to 0} Q_{2l}=\tfrac{1}{3}Bb^3\left(2-3\eta+\eta^3\right)$.

The total flow rate of fluid 2 is:

$$Q_{2u}+Q_{2l}=\tfrac{2}{3}Bb^3\left\{2-3\eta\left(1+\varepsilon^2\right)+\eta^3\right\} \tag{2.16}$$

$$\lim_{\varepsilon\to 0}\left(Q_{2u}+Q_{2l}\right)=\tfrac{2}{3}Bb^3\left(2-3\eta+\eta^3\right)$$

and its average velocity is:

$$V_2=\frac{Q_{2u}+Q_{2l}}{2b(1-\eta)} \tag{2.17}$$

Then, the global holdup ratio h_g can be defined as follows:

$$h_g=\frac{c}{V_2}=\frac{3(1+\eta)\left\{(1-\eta)^2-\varepsilon^2\right\}}{2-3\eta\left(1+\varepsilon^2\right)+\eta^3} \tag{2.18}$$

$$\lim_{\varepsilon\to 0} h_g=\frac{3(1-\eta)\left(1-\eta^2\right)}{2-3\eta+\eta^3}$$

We also can define local holdup ratios for the upper and lower halves or h_u and h_l, as follows:

$$h_u=\frac{c}{V_{2u}}=\frac{3(1+\eta)\left\{(1-\eta)^2-\varepsilon^2\right\}}{2-3\eta\left(1+\varepsilon^2\right)+\eta^3-\varepsilon(1-\eta)\left\{(1-\eta)+\varepsilon\right\}} \tag{2.19}$$

$$h_l=\frac{c}{V_{2l}}=\frac{3(1+\eta)\left\{(1-\eta)^2-\varepsilon^2\right\}}{2-3\eta\left(1+\varepsilon^2\right)+\eta^3+\varepsilon(1-\eta)\left\{(1-\eta)-\varepsilon\right\}} \tag{2.20}$$

Notice that $h_u \neq h_l \neq h_g$. However,

$$\lim_{\varepsilon\to 0} h_u=\lim_{\varepsilon\to 0} h_l=\lim_{\varepsilon\to 0} h_g=\frac{3(1-\eta)\left(1-\eta^2\right)}{2-3\eta+\eta^3}=h_{sym} \tag{2.21}$$

where h_{sym} corresponds to the holdup ratio for the symmetric case.

2.3 SYMMETRIC VS. ECCENTRIC 2D PERFECT CORE FLOW

Let $c_{sym}=c_{non-sym}$ or

$$\Gamma b^2\left(1-\eta^2\right)=Bb^2\frac{(1+\eta)}{(1-\eta)}\left\{(1-\eta)^2-\varepsilon^2\right\} \tag{2.22}$$

then

$$B = \Gamma \frac{(1-\eta)^2}{\{(1-\eta)^2 - \varepsilon^2\}}$$

where $\lim_{\varepsilon \to 0} B = \Gamma$, or

$$\beta = \gamma \frac{(1-\eta)^2}{\{(1-\eta)^2 - \varepsilon^2\}} \tag{2.23}$$

and $\lim_{\varepsilon \to 0} \beta = \gamma$

Notice that for given c, a and b; $\beta > \gamma$. The pressure drop in the non-symmetric case is higher than the pressure drop in the symmetric case. Only in the limit where $\varepsilon \to 0$ they are equal, as expected. It is worth to point out that the holdup ratio does not depend on the core's velocity c. Thus, $h_u \neq h_l \neq h_g$ for any c in the 2D non symmetric PCAF.

We can conclude that for given c, a and b, the minimum pressure drop occurs when the core is centered. Only then the global holdup ratio is equal to the local holdup ratios.

3. 2D Wavy Core Flow

Consider two eccentric immiscible fluids flowing down an infinite two dimensional horizontal pipeline; the core is occupied by fluid 1 and the annulus by fluid 2, as shown in figure 3.1.

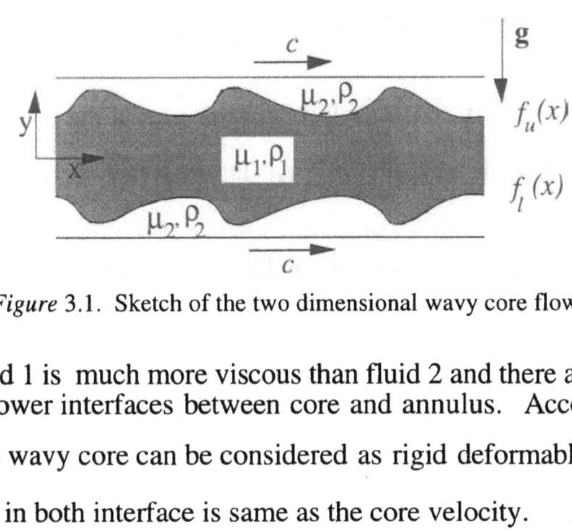

Figure 3.1. Sketch of the two dimensional wavy core flow.

Assume that fluid 1 is much more viscous than fluid 2 and there are periodic interfacial waves in upper and lower interfaces between core and annulus. According to Bai, Kelkar & Joseph [1996], the wavy core can be considered as rigid deformable body if $\frac{\mu_2}{\mu_1} < 10^{-5}$. Thus the wave speed in both interface is same as the core velocity.

3.1 PRESSURE FIELD AND GOVERNING EQUATIONS

We set up a moving frame with same core speed (see figure 3.1). We also assume that there is a connection somewhere between upper annulus and lower annuli. Thus, the pressure in the core is

$$P_1 = \beta x - \rho_1 g y + C_1 \tag{3.1.1}$$

and the pressure in both upper and lower annuli is

$$P_2 = \beta x - \rho_2 g y + p(x,y) + C_2 \tag{3.1.2}$$

The following are the governing equations for the velocity field at the indicated regions.

$$\begin{gathered}\frac{\partial u}{\partial x}+\frac{\partial v}{\partial y}=0\\ \rho_2\left(u\frac{\partial u}{\partial x}+v\frac{\partial u}{\partial y}\right)=-\beta-\frac{\partial p}{\partial x}+\mu_2\left(\frac{\partial^2 u}{\partial x^2}+\frac{\partial^2 u}{\partial y^2}\right),\\ \rho_2\left(u\frac{\partial v}{\partial x}+v\frac{\partial v}{\partial y}\right)=-\frac{\partial p}{\partial y}+\mu_2\left(\frac{\partial^2 v}{\partial x^2}+\frac{\partial^2 v}{\partial y^2}\right)\end{gathered} \tag{3.1.3}$$

in the region $-b \le y \le f_l(x)$ *and* $f_u(x) \le y \le b$,

$$\begin{gathered}u=0\\ v=0\\ \frac{dP_1}{dx}=\beta\end{gathered}$$

in the region $f_u(x) \le y \le f_l(x)$, and $u=c$ *and* $v=0$ on the boundaries or $y=\pm b$.

The general normal stress condition is

$$(-[[P]]+2H\sigma)\boldsymbol{n}+[[2\mu\boldsymbol{D}[\hat{U}]]]\bullet\boldsymbol{n}=0, \tag{3.1.4}$$

where $\hat{U}=(u,v)$, $\boldsymbol{D}[\hat{U}]=\frac{1}{2}(\nabla\hat{U}+\nabla\hat{U}^T)$, $2H$ is the sum of the principal curvatures, σ is the coefficient of interfacial tension, $\boldsymbol{n}=\boldsymbol{n}_{12}$ is the normal from liquid 1 to 2 and $[[P]]=P_1-P_2$.

According to (3.1.4) and the rigid deformable assumption, the normal stress balance at upper interface becomes

$$C_o - p(x,f_u(x)) + (\rho_2-\rho_1)gf_u(x) = \sigma\left(\frac{-f_u''}{\left(1+f_u'^2\right)^{\frac{3}{2}}}\right), \tag{3.1.5}$$

and, at the lower interface

$$C_o - p(x,f_l(x)) - (\rho_2-\rho_1)gf_l(x) = \sigma\left(\frac{-f_l''}{\left(1+f_l'^2\right)^{\frac{3}{2}}}\right), \tag{3.1.6}$$

where $C_o = C_1 - C_2$.

3.2 GLOBAL FORCE BALANCE

3.2.1 *Vertical direction.*

The total pressure on the wavy core in y-axis (see figure 3.2) is

$$F_{pressure} = \int_0^{S_l} \boldsymbol{n}_l \bullet \boldsymbol{e}_y P_2(x, f_l) ds_u - \int_0^{S_u} \boldsymbol{n}_u \bullet \boldsymbol{e}_y P_2(x, f_u) ds_u$$
$$= \int_0^z \{p_l - p_u + \rho_2 g(f_u - f_l)\} dx \qquad (3.2.1)$$

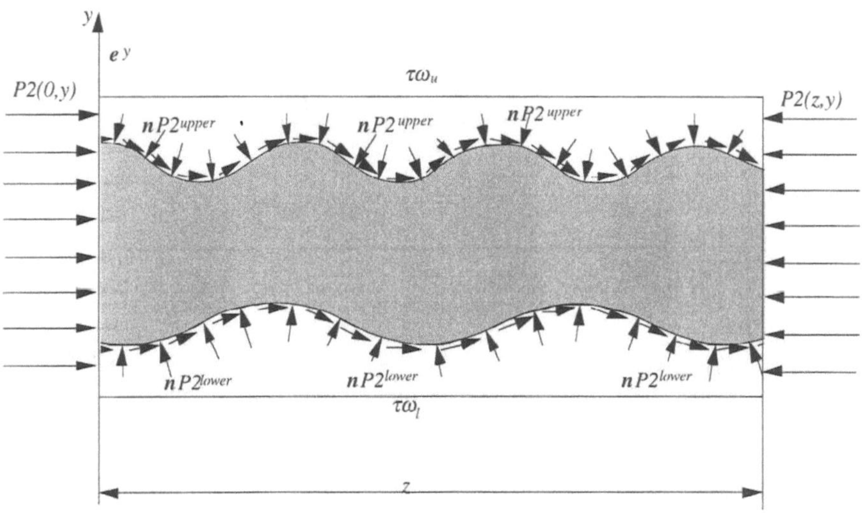

Figure 3.2. Sketch of forces acting on the core.

The total shear stress on the core in vertical direction is

$$F_{shear} = \int_0^{s_u} \boldsymbol{t}_u \tau_u \bullet \boldsymbol{e}_y ds_u - \int_0^{s_l} \boldsymbol{t}_l \tau_l \bullet \boldsymbol{e}_y ds_l$$
$$= \int_0^z \tau_u f_u' dx - \int_0^z \tau_l f_l' dx = \int_0^z (\tau_u f_u' - \tau_l f_l') dx \qquad (3.2.2)$$

The buoyancy force on the core is

$$F_{gravity} = -\int_0^z (f_u - f_l) \rho_l g dx . \qquad (3.2.3)$$

When the core is balanced, the following condition is satisfied:

$$F_{pressure} + F_{shear} + F_{bouyancy} = 0, \qquad (3.2.4)$$

or

$$\int_0^z \{p_l - p_u + \rho_2 g(f_u - f_l) + (\tau_u f_u' - \tau_l f_l') - (f_u - f_l)\rho_1 g\} dx = 0. \tag{3.2.5}$$

If we choose a very large value for z, then

$$\begin{aligned} &\lim_{z\to\infty} \frac{1}{z}\int_0^z \{p_l - p_u + \rho_2 g(f_u - f_l) + (\tau_u f_u' - \tau_l f_l') - (f_u - f_l)\rho_1 g\} dx \\ &= \frac{1}{L_l}\int_0^{L_l} (p_l - \tau_l f_l')dx - \frac{1}{L_u}\int_0^{L_u} (p_u - \tau_u f_u')dx + (\rho_2 - \rho_1)g\mathbf{2a} = 0 \end{aligned} \tag{3.2.6}$$

where L_l and L_u are the wave lengths of the lower and upper interfaces, and **2a** is the core mean size or

$$2a = \overline{f_u} - \overline{f_l} = \frac{1}{L_u}\int_0^{L_u} f_u dx - \frac{1}{L_l}\int_0^{L_l} f_l dx \tag{3.2.7}$$

Therefore the balance condition in the vertical direction is

$$\frac{1}{L_u}\int_0^{L_u} (p_u - \tau_u f_u')dx - \frac{1}{L_l}\int_0^{L_l} (p_l - \tau_l f_l')dx = (\rho_2 - \rho_1)g\mathbf{2a} \tag{3.2.8}$$

3.2.2 *Horizontal direction.*
In the x-axis, the global shear stress balance the pressure gradient as follows

$$\begin{aligned} &\int_0^z \tau_{w_u} dx + \int_0^z \tau_{w_l} dx \\ &= 2b\beta z + \int_{f_u(z)}^{D_2} p_u(y,z)dy - \int_{f_u(0)}^{D_2} p_u(y,0)dy + \int_0^{f_l(z)} p_l(y,z)dy - \int_0^{f_l(z)} p_l(y,0)dy \\ &= 2b\beta z + \Omega \end{aligned} \tag{3.2.9}$$

where ***2b*** is the gap between the parallel plates and,

$$\Omega = \int_{f_u(z)}^{D_2} p_u(y,z)dy - \int_{f_u(0)}^{D_2} p_u(y,0)dy + \int_0^{f_l(z)} p_l(y,z)dy - \int_0^{f_l(z)} p_l(y,0)dy \tag{3.2.10}$$

is bounded. Therefore,

$$\lim_{z\to\infty}\frac{1}{z}\left(\int_0^z \tau_{w_u} dx + \int_0^z \tau_{w_l} dx = 2b\beta z + \Omega\right) \text{and} \frac{1}{2b}\left(\frac{1}{L_u}\int_0^{L_u} \tau_{w_u} dx + \frac{1}{L_l}\int_0^{L_l} \tau_{w_l} dx\right) = \beta \tag{3.2.11}$$

since $\lim_{z\to\infty}\frac{\Omega}{z} \to 0.$

4. Dimensionless variables for the 2D wavy core flow

In our formulation lengths are scaled with the mean gap size δ between the core and the wall, pressures are scaled by $\rho_2 c^2$, and velocities are scaled with the wave speed $\boldsymbol{c}$, so

$$u = c\bar{u} \tag{4.1}$$

$$v = c\bar{v} \tag{4.2}$$

$$x, y, f, L = \delta\bar{x}, \delta\bar{y}, \delta\bar{f}, \delta\bar{L} \tag{4.3}$$

where $\delta = b(1-\eta)$ and $\eta = \dfrac{a}{b}$.

After equation (3.2.7) η may be expressed as :

$$\eta = \frac{a}{b} = \frac{1}{2b}\left(\frac{1}{L_u}\int_0^{L_u} f_u dx - \frac{1}{L_l}\int_0^{L_l} f_l dx\right) \tag{4.4}$$

If we split the dimensional domain (see figure 4.1) through the core's axis and change the frame of reference for the lower half as shown in figure 4.2, we can also write the following expression.

$$\frac{\eta}{1-\eta} = \frac{1}{\bar{L}}\int_0^{\bar{L}} \bar{f} d\bar{x} \tag{4.5}$$

where $\bar{L}$ *and* $\bar{f}$ stand for and $\overline{L_u}$ *or* $\overline{L_l}$ and $\overline{f_u}$ *or* $\overline{f_l}$ respectively.

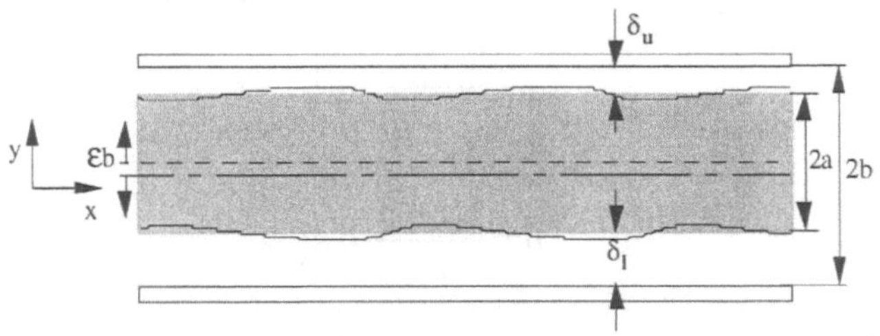

Figure 4.1 Dimensional Domain for a 2D wavy core flow.

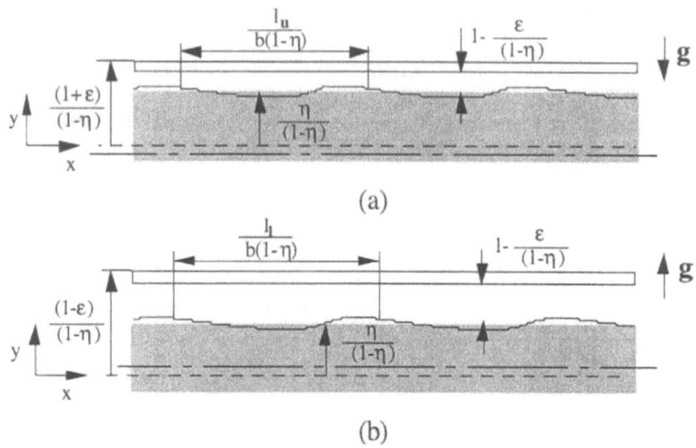

Figure 4.2 Dimensionless Domain for a 2D wavy core flow. (a) Upper half. (b) Lower half.

4.1 THE DIMENSIONLESS GOVERNING EQUATIONS

The dimensionless Navier-Stokes equations (3.1.3) are:

$$\begin{gathered}\frac{\partial \bar{u}}{\partial \bar{x}}+\frac{\partial \bar{v}}{\partial \bar{y}}=0\\ \left(\bar{u}\frac{\partial \bar{u}}{\partial \bar{x}}+\bar{v}\frac{\partial \bar{u}}{\partial \bar{y}}\right)=-\bar{\beta}-\frac{\partial \bar{p}}{\partial \bar{x}}+\frac{1}{\Re}\left(\frac{\partial^2 \bar{u}}{\partial \bar{x}^2}+\frac{\partial^2 \bar{u}}{\partial \bar{y}^2}\right),\\ \left(\bar{u}\frac{\partial \bar{v}}{\partial \bar{x}}+\bar{v}\frac{\partial \bar{v}}{\partial \bar{y}}\right)=-\frac{\partial \bar{p}}{\partial \bar{y}}+\frac{1}{\Re}\left(\frac{\partial^2 \bar{v}}{\partial \bar{x}^2}+\frac{\partial^2 \bar{v}}{\partial \bar{y}^2}\right)\end{gathered} \tag{4.6}$$

where $\Re$ is the Reynolds number and $\bar{\beta}$ is a dimensionless pressure drop, respectively. defined as:

$$\Re=\frac{\rho_2 cb(1-\eta)}{\mu_2} \tag{4.7}$$

and

$$\bar{\beta}=\frac{b(1-\eta)}{\rho_2 c}\beta \tag{4.8}$$

The boundary conditions are given by (4.9) and (4.10).

$$\bar{u}=\bar{v}=0 \qquad \text{at} \qquad \bar{y}=\bar{f}_u \text{ and } \bar{y}=\bar{f}_l \tag{4.9}$$

$$\begin{aligned}\bar{v}&=0\\ \bar{u}&=1\end{aligned} \qquad \text{at} \qquad \bar{y}=\frac{1\pm\varepsilon}{1-\eta} \tag{4.10}$$

since the dimensionless wall speed is

$$\bar{c}=\frac{c}{c}=1 \tag{4.11}$$

For the upper interface, the normal stress balance equation becomes

$$\bar{p}(\bar{x},\bar{f}_u)-\bar{C}_o=(1-\eta)\frac{J}{\Re^2}\left(\frac{(1-\eta)^2}{We}\bar{f}_u+\frac{\bar{f}_u''}{\left(1+\bar{f}_u'^2\right)^{3/2}}\right) \tag{4.12}$$

and for the lower interface becomes

$$\bar{C}_o-\bar{p}(\bar{x},\bar{f}_l)=(1-\eta)\frac{J}{\Re^2}\left(\frac{(1-\eta)^2}{We}\bar{f}_l-\frac{\bar{f}_l''}{\left(1+\bar{f}_l'^2\right)^{3/2}}\right) \tag{4.13}$$

where We is the Weber number, defined as:

$$We=\frac{\sigma}{(\rho_2-\rho_1)gb^2}, \tag{4.14}$$

and

$$J=\frac{\rho_2 b\sigma}{\mu_2^2} \tag{4.15}$$

Notice that $\frac{1}{We}\rightarrow 0$ as $(\rho_2-\rho_1)\rightarrow 0$; and equations (4.12) and (4.13) collapse into a single one, which corresponds to the axisymmetric case. Also notice that if $(\rho_2-\rho_1)<0$, equation (4.12) becomes equation (4.13) and vice versa, but the nature of the problem remains unchanged. For that reason, all our computations were carried out for $(\rho_2-\rho_1)>0$.

Finally, the force balance in the vertical and horizontal directions expressed by equations (3.2.8) and (3.2.11) respectively may be rewritten as:

$$\frac{1}{\bar{L}_u}\int_0^{\bar{L}_u}(\overline{p_u}-\overline{\tau_u}\overline{f_u}')d\bar{x}-\frac{1}{\bar{L}_l}\int_0^{\bar{L}_l}(\overline{p_l}-\overline{\tau_l}\overline{f_l}')d\bar{x}=\frac{2\eta(1-\eta)^2}{We}\frac{J}{\Re^2} \tag{4.16a}$$

$$\frac{1}{\bar{L}_u}\int_0^{\bar{L}_u}\overline{\tau_{w_u}}d\bar{x}+\frac{1}{\bar{L}_l}\int_0^{\bar{L}_l}\overline{\tau_{w_l}}d\bar{x}=2\frac{\bar{\beta}}{(1-\eta)} \tag{4.16b}$$

Therefore, only five parameters are required for a complete description of our problem: $\Re$, We, J, η and the global holdup ratio h_g which will be discussed in the next section.

4.2 THE HOLDUP RATIO

The global holdup ratio, previously defined as the ratio of the average velocities of fluid 1 and fluid 2, can be expressed as:

$$h=\frac{Q_1}{Q_2}\frac{2\delta}{2a}=c\frac{2\delta}{Q_2} \tag{4.17}$$

where Q_1 and Q_2 are the total flow rates of these fluids, δ is the mean gap size between the core and the walls and, $c = \frac{Q_1}{2a}$ is the average velocity of fluid 1.

It is easy to show that the upper and lower holdup ratios can be written as

$$h_u = c\frac{\delta_u}{Q_{2_u}}, \tag{4.18a}$$

$$h_l = c\frac{\delta_l}{Q_{2_l}}. \tag{4.18b}$$

where Q_{2u} and Q_{2l} are the fluid2 flow rates for the upper and lower halves and, δ_u and δ_l are the gap size between the core and the upper and lower walls, respectively. We also know that

$$2\delta = \delta_u + \delta_l \tag{4.19}$$

$$Q_2 = Q_{2_u} + Q_{2_l} \tag{4.20}$$

Using (4.17), (4.18a) and (4.18b), equation (4.19) can be rewritten as follows:

$$hQ_2 = h_u Q_{2_u} + h_l Q_{2_l} \tag{4.21}$$

And, the system of equations reduces to equations (4.20) and (4.21). One solution is:

$$h = h_u = h_l \tag{4.22}$$

Combining (4.17), (4.19) and (4.20), we obtain another look of the system, namely

$$\begin{aligned} h &= c\frac{\delta_u + \delta_l}{Q_{2_u} + Q_{2_l}} \\ h_u &= c\frac{\delta_u}{Q_{2_u}} \\ h_l &= c\frac{\delta_l}{Q_{2_l}} \end{aligned} \tag{4.23}$$

By inspection, we conclude that only if $h_u = h_l$, equation (4.22) holds; that is the global holdup ratio is equal to the local holdup ratio everywhere.

A more general solution is the following:

$$Q_{2_l} = \frac{h - h_u}{h_l - h_u} Q_2 \tag{4.24}$$

$$Q_{2_l} = \frac{h_l - h}{h_l - h_u} Q_2 \tag{4.25}$$

In order to keep Q_{2_u} *and* Q_{2_l} positive, we need that

$$h_l \geq h \geq h_u \quad or \quad h_u \geq h \geq h_l \tag{4.26}$$

Finally, we have to point out that the global and/or the local holdup ratios are physically restricted, thus (4.26) becomes

$$2 > h_l \geq h \geq h_u > 1 \quad or \quad 2 > h_u \geq h \geq h_l > 1 \tag{4.27}$$

The lower limit is achieved when the crests of the waves touch the wall; fluid 2 is trapped within the valleys of the interface and it travels at the same speed c. The upper limit is given for PCF when η tends to zero.

Example. Figure 4.3 shows all the possible values for the lower holdup ratio h_l as a function of the lower gap size δ_u parametrized by the upper holdup ratio h_u for $\Re$=360, η=0.76 and h=1.28. As we can see, for each upper holdup ratio h_u, the lower holdup ratio h_l may be out of bound of the region of the physical holdup ratio, except for the case of $h_l = h_u = h$.

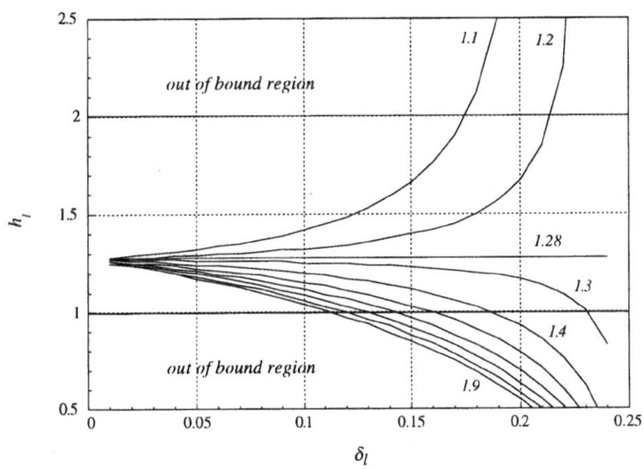

Figure 4.3. Lower holdup ratio h_l Vs. lower gap size δ_u, parametrized by the upper holdup ratio h_u for $\Re$=360, η=0.76 and h=1.28.

5. Numerical Results

Figure 5.1 presents the dimensionless pressure gradient $\overline{\beta}/(1-\eta)$ and the lower holdup ratio h_l as a function of the upper holdup ratio h_u for $\Re$=360, η=0.76, h_g=1.28, We=133 and J=1.3x10^5. It is clear that the pressure gradient is minimum when the lower and upper holdup ratios are equal, and hence, equal to the global holdup ratio or

$h_l = h_u = 1.28$. Figure 5.2 presents the eccentricity and the lower and upper wave lengths for the same case. Notice that the eccentricity also has a minimum when $h_l = h_u = 1.28$, and that the lower a upper wave length are as similar as possible. Summarizing, when the local holdup ratios are equal, the pressure gradient and the eccentricity are minimum, and the wave lengths are as close as possible, leading to a wavy core flow as symmetric (and far from the walls) as possible.

Figure 5.3 shows the dimensionless pressure gradient $\overline{\beta}$ and eccentricity ε as a function of the Reynolds number $\Re$ for η=0.80, $h_l = h_u \approx 1.29$, We=133 and J=1.3x10^5. The greater the Reynolds number, the smaller the eccentricity and the dimensionless pressure gradient $\overline{\beta}$, as we would expect, even though on dimensional grounds, the pressure gradient increases as $\Re$ increases. Figure 5.4 presents the dimensionless lower and upper wave lengths $\overline{L_1}$ and $\overline{L_2}$ as a function of the Reynolds number $\Re$ for the same case. We observe that for lower $\Re$ the upper wave length is shorter than the lower wave length, but for high $\Re$ they are almost equal.

Finally, figure 5.5 shows a sequence of cartoons that illustrate the wavy core flows corresponding to the data points that appear in figures 5.3 and 5.4. Notice how uneven the wave shapes look for the lowest Reynolds number and compare to figure 1.2, keeping in mind that this is a 3D wavy core flow. They look similar. Figures 5.3, 5.4 and 5.5 suggest that inertia helps to keep the core away from the walls and an overall flow as symmetric as possible.

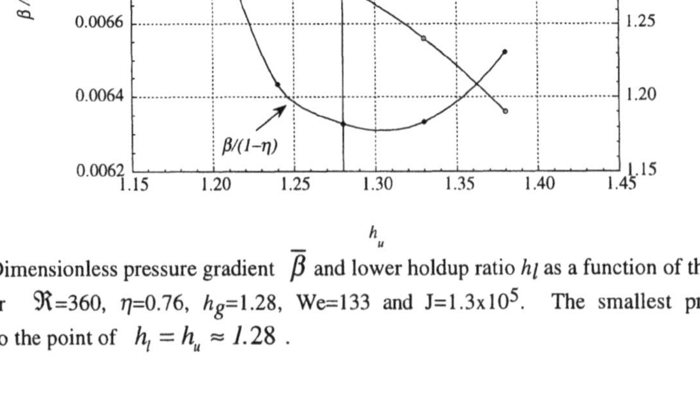

Figure 5.1. Dimensionless pressure gradient $\overline{\beta}$ and lower holdup ratio h_l as a function of the upper holdup ratio h_u, for $\Re$=360, η=0.76, h_g=1.28, We=133 and J=1.3x10^5. The smallest pressure gradient corresponds to the point of $h_l = h_u \approx 1.28$.

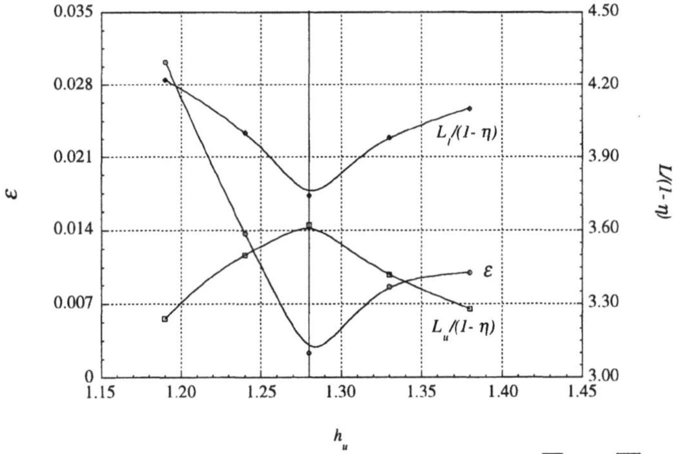

Figure 5.2. Eccentricity ε and lower and upper dimensionless wave lengths $\overline{L_l}$ and $\overline{L_u}$ as a function of upper holdup ratio h_u, for $\Re$=360, η=0.76, h_g=1.28, We=133 and J=1.3x10^5. The smallest pressure gradient corresponds to the point of $h_l = h_u \approx 1.28$.

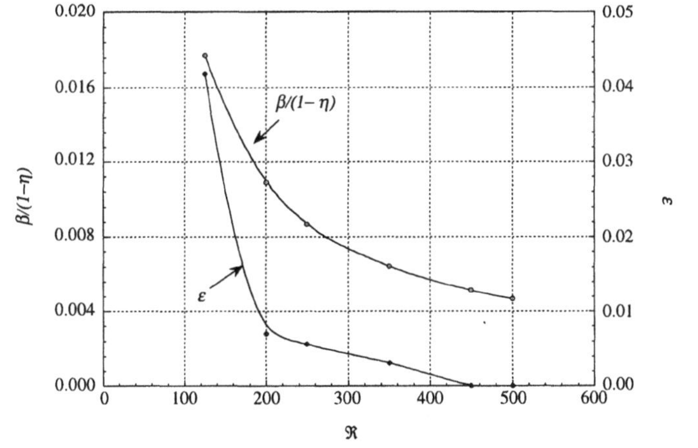

Figure 5.3. Dimensionless pressure gradient $\overline{\beta}$ and eccentricity ε as a function of the Reynolds number $\Re$, for η=0.80, $h_l = h_u \approx 1.29$, We=133 and J=1.3x10^5.

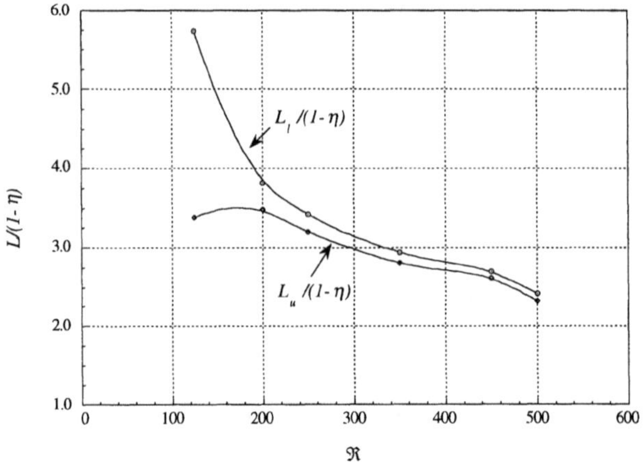

Figure 5.4. Lower and upper dimensionless wave lengths $\overline{L_l}$ and $\overline{L_u}$ as a function of Reynolds number $\Re$ for η=0.80, $h_l = h_u \approx 1.29$, We=133 and J=1.3x10^5.

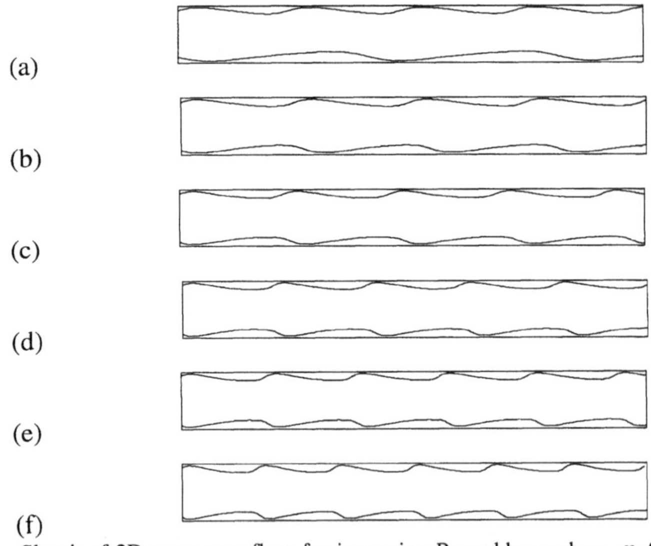

Figure 5.5. Sketch of 2D wavy core flow for increasing Reynolds numbers, η=0.80, $h_l = h_u = h_g$, We=133 and J=1.3x10^5. (a) $\Re$=125, (b) $\Re$=250, (c) $\Re$=120, (d) $\Re$=350, (e) $\Re$=450, (b) $\Re$=500.

6. Concluding remarks

Two local holdup ratios can be defined in Eccentric Perfect Core Flow (EPCF). In general, these holdup ratios are different. Only when the core is centered, they are equal and, the pressure gradient is minimum.

In 2D wavy core flows pressure gradient and eccentricity are funtion of a five-parameter set, namely $(\Re, We, J, \eta, h_g)$. Among these parameters we find the global holdup ratio. Two local holdup ratios can also be defined. When they are equal and, thus, equal to the global holdup ratio, the pressure gradient and the eccentricity are minimum. Also, there is evidence that inertia plays a role in keeping the core away from the walls.

Additional numerical work is required to completely describe the behavior of levitated core flows.

7. References

Bai, R., Kelkar, K. & Joseph, D. D. (1996) Direct simulation of interfacial waves in a high-viscosity-ratio and axisymmetric core-annular flow, *J. Fluid Mech.* **327**, 1-34.

Feng, J., Huang, P. Y., & Joseph, D. D. (1995) Dynamic simulation of the motion of capsules in pipelines, *J. Fluid Mech.* **286**, 201-227.

Joseph, D. D., & Renardy, Y. Y. (1993) *Fundamentals of two-fluid dynamics*, Springer-Verlag New York.

Oliemans, R. V. A. & Ooms, G. (1986) Core-annular flow of oil and water through a pipeline, *Multiphase Science and Technology* **2**, eds. Hewitt, G. F., Delhaye, J. M. & Zuber, N., Hemisphere Publishing Corporation.

Ooms, G., Segal, A., Van der Wees, A. J., Meerhoff, R. & Oliemans, R. V. A. (1984) A theoretical model for core-annular flow of a very viscous oil core and a water annulus through a horizontal pipe, *Int. J. Multiphase Flow* **10**, 41-60.

Patankar, S. (1980) *Numerical heat transfer and fluid flow*, eds. Minkowycz, W. & Sparrow, M., Hemisphere Publishing Corporation.

Appendix - Computational Solution of the 2D Wavy Core Flow

The two dimensional wavy core flow is governed by equations (4.4) to (4.7); the normal stress condition on the upper and lower surfaces is specified in equations (4.8) and (4.9); and the force balance on the oil core for vertical and horizontal directions is described by equations (4.14) and (4.15) respectively. For each given value of the dimensionless five-parameter set $(\Re, We, J, \eta, h_g)$, computational solution of these equations is carried out to determine the flow of water, the shape and location of the upper and lower interfaces, their wavelengths, and the eccentricity. This task is iteratively achieved in two steps. First, the computational domain is split at the center of the core; then iterative solutions of the water flow field and the interface shape are found for each half. Secondly, the net force on the vertical direction is computed and if not zero, the location of the core center is modified in order to correct it. In this step we also compute the total shear stress acting on the upper and lower walls and the pressure gradient $\overline{\beta}$ is updated. The overall solution method involves an outer iteration related to the balance of the vertical forces acting on the core and a two-step inner iteration related to the calculation of the water flow field and determination of the upper and lower interfaces. The procedure is outlined below.

1. Prescribe the wave speed c, properties of fluids 1 and 2, such as density jump $(\rho_2-\rho_1)$, surface tension σ and viscosity μ_2, values for the core size ***2a*** and walls separation ***2b***

and the global holdup ratio h_g, which corresponds to the dimensionless five-parameter set $(\Re, We, J, \eta, h_g)$.

2. Assume the upper and lower interfaces shape. For each half of the domain, calculate the velocity and pressure fields in the water region for the specified Reynolds Number $\Re$. The Patankar's finite control volume method is used (1980). During each iteration of the flow field calculation, the pressure gradient $\overline{\beta}$ is adjusted to satisfy the force balance in the horizontal direction on the oil plug.
3. The shape of each interface is determined by solving the appropriate equation describing the normal stress condition for the interfacial pressure distribution determined in step 2. The wave length $\overline{L}$ and pressure constant $\overline{C}_o$ for each interface are adjusted in each iteration so that at convergence each interface shape is determined for the prescribed value of η and the difference between the available local holdup ratios.
4. Calculate the net vertical force and according to its sign move the location of the center of the core upwards or downwards, keeping track of the displacement.
5. The new upper and lower interfaces and the new location of the center of the core, which is a measure of the eccentricity, are used now in determining the flow field in step 2. Thus, steps 2, 3 and 4 are repeated till the vertical forces are balanced on the core. When it happens, a self-consistent flow field of water is obtained; in particular, the pressure constant $\overline{C}_o$ is the same for the lower and upper halves and the local holdup ratios are equal to the global holdup ratio. Also are obtained the shape of the upper and lower interfaces and the eccentricity of the core for the prescribed values of the five-parameter set $(\Re, We, J, \eta, h_g)$

This overall method was adapted from the one used by Bai et al [1996], which was proven to predict the same interface shape in the flow field of water irrespective of the initial guess interface. Consequently, the above method was applied for computing the details of the 2D wavy core annular flow for a range of the five-parameter set $(\Re, We, J, \eta, h_g)$.

AXISYMMETRIC WAVY CORE FLOW IN A HEAVY VISCOUS OIL

R. BAI[1], D. D. JOSEPH[1] and K. KELKAR[2]
[1] *Department of Aerospace Engineering and Mechanics*
University of Minnesota
110 Union ST. SE.
Minneapolis, MN 55455, USA

[2] *Innovative Research, Inc.*
2800 University Ave. SE.
Minneapolis, MN 55414, USA

Abstract

A direct numerical simulation of spatially periodic wavy core flows is carried out under the assumption that the densities of the two fluids are identical and that the viscosity of the oil core is so large that it moves as a rigid solid which may nevertheless be deformed by pressure forces in the water. The waves which develop are asymmetric with steep slopes in the high pressure region at the front face of the wave crest and shallower slopes at low pressure region at lee side of the crest. The simulation gives excellent agreement with the experiments of Bai et al.[1992] on up flow in vertical core flow where axisymmetric bamboo waves are observed. We define a threshold Reynolds number and explore its utility; the pressure force of the water on the core relative to a fixed reference pressure is negative for Reynolds numbers below the threshold and is positive above. The wave length increases with the holdup ratio when the Reynolds number is smaller than a second threshold and decreases for larger Reynolds numbers. We verify that very high pressures are generated at stagnation points on the wave front. It is suggested that the positive pressure force is required to levitate the core off the wall when the densities are not matched and to center the core when they are. A further conjecture is that the principal features which govern wavy core flows cannot be obtained from any theory in which inertia is neglected.

---

1. Introduction

Lubricated pipelining of viscous materials like heavy crude, slurries and capsules is robustly stable and has a high economic potential. The viscous material does not touch the wall. In the case of crude oil, the drag reduction which can be achieved by lubrication is of the order of the viscosity ratio with increased throughputs of ten thousand or more (for more background see Joseph and Renardy 1992). These lubricated flows are called core flows because the viscous material flows in a core lubricated all around by water.

Probably the most important industrial pipeline to date was the 6-inch (15.2 cm) diameter, 24-mile (38.6 km) long Shell line from the North Midway Sunset Reservoir near Bakersfield, California, to the central facilities at Ten Section. The line was run under the supervision of Veet Kruka for 12 years from 1970 until the Ten section was closed. When lubricated by water at a volume flow rate of 30% of the total, the pressure drop varied between 900 psi and 1,100 psi at a flow rate of 24,000 barrels per day with the larger pressure at a threshold of unacceptability which called for pigging. In the sixth year of operation the fresh water was replaced with water produced at the well site which contained various natural chemicals leached from the reservoir, including sodium metasilicate in minute 0.6 wt. % amounts. After that the pressure drop never varied much from the acceptable 900 psi value; the CAF was stable as long as the flow velocity was at least 3 ft/s.

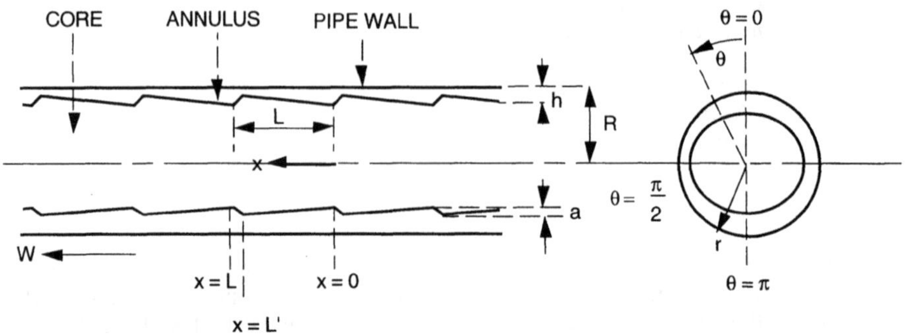

Figure 1.1. (after Ooms et.al., *Int. J. Multiphase Flow*, **10**, 41-60, 1984). The core is at rest and the pipe wall moves to the left.

A surprising property of core flow is that the flow in a horizontal line will lubricate with the core levitated off the wall even if the core is lighter or heavier than lubricating water. This levitation could not take place without a hydrodynamic lifting action due to waves sculpted on the core surface. In the case of very viscous liquids, the waves are basically standing waves which are convected with the core as it moves downstream. This picture suggests a lubrication mechanism for the levitation of the core analogous to mechanisms which levitate loaded slider leavings at low Reynolds numbers. Ooms, Segal, Vander Wees, Meerhoff and Oliemans [1984] and Oliemans and Ooms [1986] gave a semi-emperical model of this type and showed that it generated buoyant forces proportional to the first power of the velocity to balance gravity. In this theory, the shape of the wave must be given as an empirical input (see Figure 1.1).

Consider water lubricated pipelining of crude oil. The oil rises up against the pipe wall because it is lighter than the water. It continues to flow because it is lubricated by waves. However, the conventional mechanisms of lubrication cannot work. The saw tooth waves shown in Figure 1.1 are like an array of slipper bearings and the stationary oil core is pushed off the top wall by lubrication forces. If W were reversed, the core would be sucked into the wall, so the slipper bearing picture is obligatory if you want levitation.

Obviously the saw tooth waves are unstable since the pressure is highest just where the gap is smallest, so the wave must steepen where it was gentle, and smooth where it was sharp. This leads us to the cartoon in Figure 1.2. To get a lift from this kind of wave it appears that we need inertia. Liu's [1982] formula for capsule lift-off in a pipeline in which the critical lift off velocity is proportional to the square root of gravity times the density difference is an inertial criterion. Feng, Huang & Joseph [1995] did direct numerical simulation of the transport of elliptical shaped capsules in pipelines and found that lubrication and initial effects coexist; the lift due to lubrication is dominant for low flying capsules and inertia is dominant for high flying capsules. It is likely that similar dynamics are involved in lubricated oil and slurry lines. At high speeds the core flows may literally "fly" down the tube. In all of this, the position of the viscous points of stagnation where the pressures are high is of critical importance.

The importance of inertia in levitation of core flows is also suggested by industrial experience in which stable lubricated flow occurs only when the flow velocity exceed some critical value.

In this paper we confine our attention to the direct numerical simulation of axisymmetric core flows. The shape of the interface and the secondary motions which develop in a "flying" core flow are consistent with the picture arising intuition (see Figure 1.2).

Less intuitive is the existence of a threshold Reynolds number corresponding to a change in the sign of the pressure force on the core, from suction at Reynolds numbers below the threshold, as in the reversed slipper bearing in which the slipper is sucked to the wall, to compression for Reynolds numbers greater than the threshold as in flying core flow in which the core can be pushed off the wall by stagnation pressure.

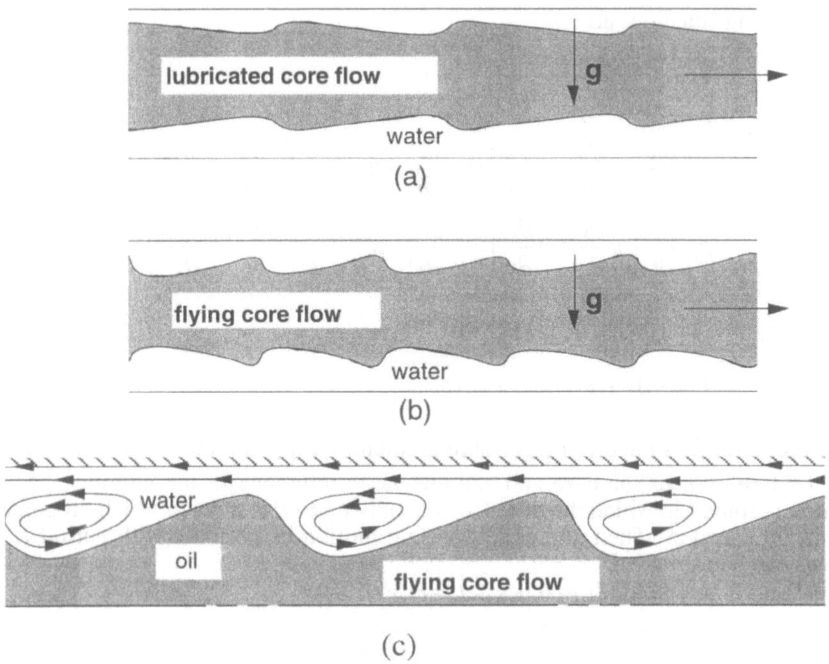

Figure 1.2. (After Feng, Huang & Joseph [1995]) (a) The interface resembles as slipper bearing with the gentle slope propagating into the water. (b) The high pressure at the front of the wave crest steepens the interface and the low pressure at the back makes the interface less steep. (c) The pressure distribution in the trough drives one eddy in each trough.

2. Equations of Motion for Two Immiscible Fluids with Interfacial Waves

Consider two concentric immiscible fluids flowing down an infinite horizontal pipeline; the core is occupied by fluid 1 and the annulus by fluid 2. Assume also that the core is axisymmetric with interfacial waves. The waves are periodic along the axis and move at certain uniform speed c (see Figure 2.1).

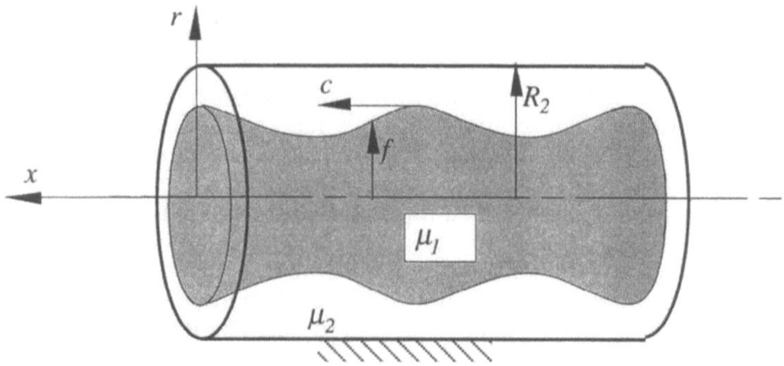

Figure 2.1. Sketch of wavy core flow.

The governing equations are

$$\frac{1}{r}\frac{\partial}{\partial r}(ru)+\frac{\partial v}{\partial x}=0,$$
$$\rho_i\left(\frac{\partial u}{\partial t}+u\frac{\partial u}{\partial r}+v\frac{\partial u}{\partial x}\right)=-\frac{\partial P}{\partial r}+\mu_i\left(\frac{1}{r}\frac{\partial}{\partial r}\left(r\frac{\partial u}{\partial r}\right)+\frac{\partial^2 u}{\partial x^2}-\frac{u}{r^2}\right), \tag{2.1}$$
$$\rho_i\left(\frac{\partial v}{\partial t}+u\frac{\partial v}{\partial r}+v\frac{\partial v}{\partial x}\right)=-\frac{\partial P}{\partial x}+\mu_i\left(\frac{1}{r}\frac{\partial}{\partial r}\left(r\frac{\partial v}{\partial r}\right)+\frac{\partial^2 v}{\partial x^2}\right),$$

where (u,v) are the (radial, axial) velocity and

$$P=-\beta x+p(r,x,t)+C_o(t) \tag{2.2}$$

is the pressure, β is the driving pressure gradient, (μ_i,ρ_i) are the (viscosity, density), the core $i=1$ references the core and annulus $i=2$ reference to the annulus. A periodic solution is sought in which the time t enters only as $x\text{-}ct$ in a wave which propagates with a constant speed c. Then, the condition that motion be periodic with period L in x as follows can be expressed:

$$\begin{aligned} u&=u(x-ct,r)=u(x+L-ct,r)\\ v&=v(x-ct,r)=v(x+L-ct,r)\\ p(x-ct,r)&=p(x+L-ct,r). \end{aligned} \tag{2.3}$$

Next new variables are introduced as such

$$z=x-ct \tag{2.4}$$

and

$$\begin{aligned} w(z,r)&=v(x-ct,r)-c,\\ u(z,r)&=u(x-ct,r),\\ P(z,r)&=P(x-ct,r)=-\beta(z+ct)+p(r,z)+C_o(t). \end{aligned} \tag{2.5}$$

The system of equations arising from (2.1), (2.4) and (2.5) is to be time independent

$$C_o(t)-\beta ct=C_{pi}$$

where C_{pi} is constant. In these new variables, z replaces x in (2.1) and

$$\begin{aligned}
&\frac{1}{r}\frac{\partial}{\partial r}(ru)+\frac{\partial w}{\partial z}=0, \\
&\rho_i\left(u\frac{\partial u}{\partial r}+w\frac{\partial w}{\partial x}\right)=-\frac{\partial p}{\partial r}+\mu_i\left(\frac{1}{r}\frac{\partial}{\partial r}\left(r\frac{\partial u}{\partial r}\right)+\frac{\partial^2 u}{\partial z^2}-\frac{u}{r^2}\right), \\
&\rho_i\left(u\frac{\partial w}{\partial r}+w\frac{\partial w}{\partial z}\right)=\beta-\frac{\partial p}{\partial z}+\mu_i\left(\frac{1}{r}\frac{\partial}{\partial r}\left(r\frac{\partial w}{\partial r}\right)+\frac{\partial^2 w}{\partial z^2}\right).
\end{aligned} \tag{2.6}$$

At the pipe wall $r = R_2$

$$u=0, \qquad w=-c, \tag{2.7}$$

and at the center of the core $r = 0$, we require

$$u=0, \qquad \frac{\partial w}{\partial z}=0. \tag{2.8}$$

The interface is given by

$$r = f(z) = f(z+L). \tag{2.9}$$

Here, between the water and oil, the kinetic conditions are

$$u=\frac{df}{dt}=\frac{\partial f}{\partial t}+v\frac{\partial f}{\partial x}=w\frac{\partial f}{\partial z}, \tag{2.10}$$

and the velocity condition is $[[\hat{\boldsymbol{U}}]]=0$, where $[[\bullet]]=(\bullet)_1-(\bullet)_2$ and $\hat{\boldsymbol{U}}=(u,w)$. The traction vector balancing with surface tension on the interface is derived by Joseph and Renardy [1992] as

$$(-[[P]]+2H\sigma)\boldsymbol{n}+[[2\mu\boldsymbol{D}[\hat{U}]]]\bullet\boldsymbol{n}=0. \tag{2.11}$$

Then the normal stress condition is

$$(-[[P]]+2H\sigma)+\boldsymbol{n}\bullet[[2\mu\boldsymbol{D}[\hat{U}]]]\bullet\boldsymbol{n}=0, \tag{2.12}$$

and the shear stress condition is

$$\boldsymbol{t}\bullet[[2\mu\boldsymbol{D}[\hat{U}]]]\bullet\boldsymbol{n}=0, \tag{2.13}$$

where $\boldsymbol{D}[\boldsymbol{U}]=\frac{1}{2}(\nabla\hat{\boldsymbol{U}}+\nabla\hat{\boldsymbol{U}}^{T})$, $2H$ is the sum of the principal curvatures, σ is the coefficient of interfacial tension, $\boldsymbol{n}=\boldsymbol{n}_{12}$ is the normal from liquid 1 to 2, and $\boldsymbol{t}$ is the tangent vector.

In addition, we prescribe the oil flow rate

$$Q_o=\frac{1}{L}\int_0^L\{\int_0^f 2\pi r v(r,z)dr\}dz, \tag{2.14}$$

and water flow rate

$$Q_w=\frac{1}{L}\int_0^L\{\int_f^{R_2} 2\pi r v(r,z)dr\}dz. \tag{2.15}$$

3. The Interfacial Equations

The interface is given by

$$r=f(z). \tag{3.1}$$

The unit tangent vector is given by

$$\boldsymbol{t}=\frac{(df/dz)\boldsymbol{e}_r+\boldsymbol{e}_z}{\sqrt{1+(df/dz)^2}}=\frac{f'\boldsymbol{e}_r+\boldsymbol{e}_z}{\sqrt{1+f'^2}}=\frac{1}{\sqrt{1+f'^2}}(f',1). \tag{3.2}$$

The unit normal vector is given by

$$\boldsymbol{n}=\frac{\boldsymbol{e}_r-(df/dz)\boldsymbol{e}_e}{\sqrt{1+(df/dz)^2}}=\frac{\boldsymbol{e}_r-f'\boldsymbol{e}_z}{\sqrt{1+f'^2}}=\frac{1}{\sqrt{1+f'^2}}(1,-f'). \tag{3.3}$$

From $u=w\frac{\partial f}{\partial z}$ in the interface, there is $u-w\frac{df}{dz}=0$, and therefore

$$0=\frac{u-w(df/dz)}{\sqrt{1+(df/dz)^2}}=(u\boldsymbol{e}_r+w\boldsymbol{e}_z)\bullet\left(\frac{\boldsymbol{e}_r-(df/dz)\boldsymbol{e}_z}{\sqrt{1+(df/dz)^2}}\right)=\hat{\boldsymbol{U}}\bullet\boldsymbol{n},$$

the velocity normal to the interface is always zero.

The mean curvature is given by

$$2H=\frac{1}{f\sqrt{1+(df/dz)^2}}-\frac{d^2f/dz^2}{(1+(df/dz)^2)^{1.5}}. \tag{3.4}$$

The stress tensor is given by

$$\mathbf{D}[\hat{U}] = \frac{1}{2}(\nabla\hat{U} + \nabla\hat{U}^T) = \begin{bmatrix} 2\dfrac{\partial w}{\partial z} & \dfrac{\partial w}{\partial r} + \dfrac{\partial u}{\partial z} \\ \dfrac{\partial w}{\partial r} + \dfrac{\partial u}{\partial z} & 2\dfrac{\partial u}{\partial r} \end{bmatrix} \tag{3.5}$$

where $\hat{U} = u\boldsymbol{e}_r + w\boldsymbol{e}_z = u(r,z)\boldsymbol{e}_r + w(r,z)\boldsymbol{e}_z$,

Therefore, the equation (2.13) for shear stress is

$$[[\mu\left(2\frac{df}{dz}\left(\frac{\partial u}{\partial r} - \frac{\partial w}{\partial z}\right) + \left(1 - \left(\frac{df}{dz}\right)^2\right)\left(\frac{\partial w}{\partial r} + \frac{\partial u}{\partial z}\right)\right)]] = 0. \tag{3.6}$$

The general equation for normal stress is

$$\frac{2}{1+\left(\dfrac{df}{dz}\right)^2}[[\mu\left(\left(\frac{df}{dz}\right)^2\frac{\partial w}{\partial z} - \frac{df}{dz}\left(\frac{\partial w}{\partial r} + \frac{\partial u}{\partial z}\right) + \frac{\partial u}{\partial r}\right)]] - ([[p(z,f)]] - 2H\sigma) = 0, \tag{3.7}$$

or

$$\frac{2}{1+\left(\dfrac{df}{dz}\right)^2}[[\mu\left(\left(\frac{df}{dz}\right)^2\frac{\partial w}{\partial z} - \frac{df}{dz}\left(\frac{\partial w}{\partial r} + \frac{\partial u}{\partial z}\right) + \frac{\partial u}{\partial r}\right)]] - \left([[p(z,f)]] - \frac{\sigma}{f\sqrt{1+\left(\dfrac{df}{dz}\right)^2}} + \frac{\sigma\dfrac{d^2f}{dz^2}}{\left(1+\left(\dfrac{df}{dz}\right)^2\right)^{3/2}}\right) = 0. \tag{3.8}$$

4. Rigid Deformable Core Flow

In many situations, the viscosity of the oil is much greater than the viscosity of water. We shall assume that the core is solid with standing waves on the interface. A consequence of this assumption is that non-rigid motions of the core may be neglected. In this rigid deformable core model, the core is stationary ($\hat{\boldsymbol{U}}_I = 0$) and water moves. The velocity for wavy core flow can be written (Figure 4.1) as

$$\hat{\boldsymbol{U}} = u\boldsymbol{e}_r + w\boldsymbol{e}_z = U_n\boldsymbol{n} + U_s\boldsymbol{t} \tag{4.1}$$

where U_n is normal component and U_s is tangential component of the velocity on the interface $r=f(z)$.

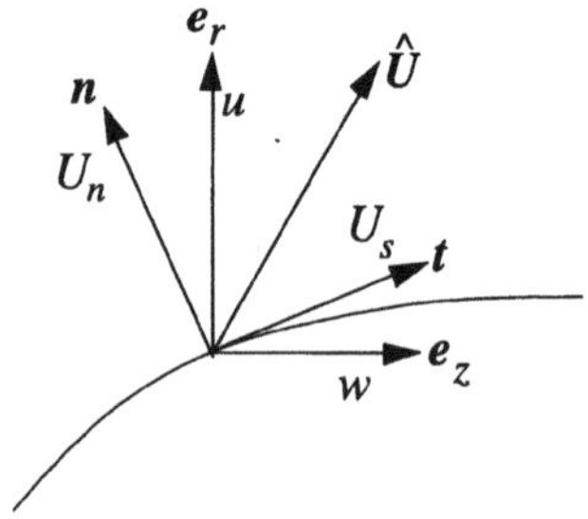

Figure 4.1. The velocity components are in different coordinates.

The continuity of the shear stress gives rise to

$$\frac{\partial U_{s1}}{\partial n} = \frac{\mu_2}{\mu_1}\frac{\partial U_{s2}}{\partial n} = m\frac{\partial U_{s2}}{\partial n}. \tag{4.2}$$

So, no matter what is the value $\frac{\partial U_{s2}}{\partial n} = O(1)$, $\frac{\partial U_{s1}}{\partial n} \sim O(m)$.

For heavy crudes in water $m = \frac{\mu_2}{\mu_1} = O(10^{-5})$ is very small and compatible with the assumption that $\hat{\boldsymbol{U}}_1$ is effectively zero.

From continuity of velocity on the interface, $[[\hat{U}]] = 0$, we also obtain that the water velocity is zero on the interface. Since the relative velocity in the solid core vanishes, the speed c of the advected wave in coordinates in which the wall is stationary and the core moves forward is exactly the core velocity $c=w(r,z)$ which is constant, independent of r and z when $r \le f(z)$. In this case (2.14) reduces to

$$Q_o = \pi R_1^2 c, \tag{4.3}$$

and

$$R_1^2 = \frac{1}{L}\int_0^L f^2(z)dz. \tag{4.4}$$

The viscous part of the normal stress on the interface vanishes:

$$\boldsymbol{n} \bullet [[2\mu \boldsymbol{D}[\hat{U}]]] \bullet \boldsymbol{n} = 2\mu_1 \frac{\partial U_{n1}}{\partial n} - 2\mu_2 \frac{\partial U_{n2}}{\partial n} = 0 \tag{4.5}$$

since

$$\frac{\partial U_s}{\partial s} + \frac{\partial U_n}{\partial n} = 0$$

and on the interface

$$\frac{\partial U_S}{\partial s} = 0. \tag{4.6}$$

Therefore, the normal stress balance (2.12) reduces to

$$[[P]] = 2H\sigma. \tag{4.7}$$

Although the core is assumed to be a deformable rigid body, there is a mass flow of oil in the pipeline, which is driven by pressure βz. The pressure in the core is

$$P_1 = -\beta z + C_{p1}, \tag{4.8}$$

while the pressure in the annulus is

$$P_2 = -\beta z + p(r,z) + C_{p2}. \tag{4.9}$$

Since the pressure level is indeterminate we may, without loss of generality, put

$$C_{p2} = 0. \tag{4.10}$$

In the water, $f \le r \le R_2$, we have

$$\begin{aligned}
&\frac{1}{r}\frac{\partial}{\partial r}(ru) + \frac{\partial w}{\partial z} = 0 \\
&\rho_2\left(u\frac{\partial u}{\partial r} + w\frac{\partial u}{\partial z}\right) = -\frac{\partial p}{\partial r} + \mu_2\left(\frac{1}{r}\frac{\partial}{\partial r}\left(r\frac{\partial u}{\partial r}\right) + \frac{\partial^2 u}{\partial z^2} - \frac{u}{r^2}\right) \\
&\rho_2\left(u\frac{\partial w}{\partial r} + w\frac{\partial w}{\partial z}\right) = \beta - \frac{\partial p}{\partial z} + \mu_2\left(\frac{1}{r}\frac{\partial}{\partial r}\left(r\frac{\partial w}{\partial r}\right) + \frac{\partial^2 w}{\partial z^2}\right).
\end{aligned} \tag{4.11}$$

And in the oil core, $0 < r < f$

$$\frac{dP_1}{dz} = -\beta. \tag{4.12}$$

The normal stress balance at $r = f$ which is given by (4.8), then becomes

$$\frac{\sigma}{f\sqrt{1+\left(\dfrac{df}{dz}\right)^2}} - \frac{\sigma\dfrac{d^2 f}{dz^2}}{\left(1+\left(\dfrac{df}{dz}\right)^2\right)^{3/2}} = [[P]] = P_1 - P_2 = C - p(z, f(z)). \tag{4.13}$$

To solve our problem we must prescribe β, and compute c, or we may give c and compute β. Since the momentum change in one wavelength is zero, we may relate c and β by the force balance shown in Figure 4.2 and expressed as

$$\begin{aligned}
A\beta L = F_z &= 2\pi\int_0^{S_L} f(\boldsymbol{n}P_2 + \boldsymbol{t}\tau) \bullet \boldsymbol{e}_z ds = 2\pi\int_0^L (\boldsymbol{n}P_2 + \boldsymbol{t}\tau) \bullet \boldsymbol{e}_z f\sqrt{1+(df/dz)^2}\,dz, \\
&= 2\pi\int_0^L (P_2 sin\theta + \tau cos\theta) f\sqrt{1+(df/dz)^2}\,dz
\end{aligned} \tag{4.14}$$

where $$tan\theta = \frac{df}{dz},\tag{4.15}$$

and βL is the pressure drop, A is the area of the cross section of the core, S_L is the arc length in one wavelength $r=f$, $\boldsymbol{n}$ is the unit normal vector, $\boldsymbol{t}$ is the unit tangent vector, $\boldsymbol{e}_z$ is the unit vector in the axial direction, and τ is the shear stress.

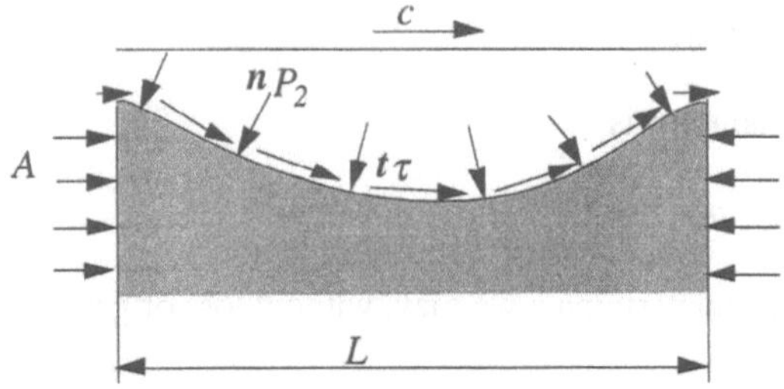

Figure 4.2. Force balance on the wave core is based on one wavelength. The pressure drop is βL and $A\beta L = F_z$ is the force.

The pressure P_2 and shear stress τ depend on c implicitly; we give c and compute P_2 and τ. The corresponding β is given by iteration using (4.14).

We must now solve (4.11) for L periodic functions u, w, p subject to the conditions that $w=c$, at $r=R_2$, $w=u=0$ at $r=f(z)$ where Q_W is prescribed by (4.15), the pressure and interfacial tension are related by (4.13) and β and c are related by (4.14). To solve this, we first prescribe the shape $f_0(z)$ of the wave with a prescribed mean R_1 and, of course, its length L_0 and the speed c. We next construct a sequence of iterations in which c, R_1, and Q_W are fixed. We first compute u, w and p satisfying all the conditions except (4.13). The pressure p_0, belonging to $f_0(z)$, then determines a new shape $f_1(z)$ to satisfy (4.13). A certain constant C_{p1} is required to maintain the value of R_1 in the iteration. At the next stage we repeat the calculation of u_1, w_1 and p_1. This gives a new $f_2(z)$ and C_{p2} and so on.

We can, therefore, calculate convergent solutions corresponding to each and every triplet (c, R_1, Q_w). We do not expect all these solutions to be stable. In experiments we get one R_1 when we prescribe the inputs Q_W and Q_O or c. This should be equivalent to prescribing two other independent bits of data, say c and R_1. Then we would expect one Q_W for each c and R_1 to appear in experiments, but we calculate a family of solutions for the given c and R_1 and any Q_W. When the parameters are taken for perfect core flow, we get perfect core flow from our simulation; most of these perfect core flows are actually unstable. Our simulation then can be regarded as giving rise to a one parameter family of wavy core flows whose stability is yet to be tested.

5. The Holdup Ratio in Wavy Core Flow

The holdup ratio h is the ratio Q_o/Q_W of volume flow rates to the ratio V_o/V_W of volume in the pipe:

$$h = \frac{Q_o / Q_W}{V_o / V_w} = \frac{Q_o/Q_W}{R_1^2/(R_2^2 - R_1^2)},\tag{5.1}$$

The holdup ratio depends on the fluid properties and flow parameters but is most strongly influenced by flow type. Equation (5.1) can be represented as the ratio of superficial velocities

$$h = \frac{Q_o/\pi R_1^2}{Q_w/\pi(R_2^2 - R_1^2)} = \frac{c_0}{c_w}, \tag{5.2}$$

where
$$c_o = \frac{Q_o}{\pi R_1^2}, \tag{5.3}$$

$$c_w = \frac{Q_w}{\pi(R_2^2 - R_1^2)}, \tag{5.4}$$

and $c = c_O$ for rigid core flow.

Here, we consider the holdup ratio in perfect core flow and wavy core flow. A perfect core annular flow has a perfectly cylindrical interface of uniform radius without waves and is perfectly centered on the pipe axis with an annulus of lubricating water all around. The velocity W and flow rate Q of a perfect core flow is

$$W_o(r) = \frac{P'}{4\mu_1}\left(R_1^2 - r^2\right) + \frac{P'}{4\mu_2}\left(R_2^2 - R_1^2\right), \tag{5.5}$$

$$W_w(r) = \frac{P'}{4\mu_2}\left(R_2^2 - r^2\right), \tag{5.6}$$

$$Q_o = 2\pi\int_0^{R_1} rW_o dr = 2\pi\left\{\frac{P'}{16\mu_1}R_1^4 + \frac{P'}{8\mu_2}(R_2^2R_1^2 - R_1^4)\right\}, \tag{5.7}$$

$$Q_w = 2\pi\int_{R_1}^{R_2} rW_w(r)dr = 2\pi\left\{\frac{P'}{16\mu_2}\left(R_2^2 - R_1^2\right)^2\right\}. \tag{5.8}$$

(5.7) and (5.8) can be written as

$$Q_o = -\frac{P'\pi R_1^4}{8\mu_2}\left(m + 2\left(a^2 - 1\right)\right), \tag{5.9}$$

and
$$Q_w = -\frac{P'\pi R_1^4}{8\mu_2}\left(a^2 - 1\right)^2. \tag{5.10}$$

The ratio of the flow rates, the input ratio, is

$$\gamma = \frac{Q_o}{Q_w} = \frac{m + 2(a^2 - 1)}{(a^2 - 1)^2}, \tag{5.11}$$

where
$$m = \frac{\mu_2}{\mu_1}, \quad a = \frac{R_2}{R_1} = \sqrt{1 + \frac{1}{\gamma}(1 + \sqrt{1 + m\gamma})}.$$

The oil fraction is given by

$$\eta^2 = \left(\frac{R_1}{R_2}\right)^2 = \frac{1}{a^2} = \frac{1}{1+\frac{1}{\gamma}\left(1+\sqrt{1+m\gamma}\right)}, \tag{5.12}$$

while the water fraction is given by

$$1-\eta^2 = 1-\frac{1}{1+\frac{1}{\gamma}\left(1+\sqrt{1+m\gamma}\right)} = \frac{1+\sqrt{1+m\gamma}}{\gamma+1+\sqrt{1+m\gamma}}. \tag{5.13}$$

The volume ratio is proportional to the fraction ratio. Hence

$$\frac{V_0}{V_w} = \frac{\eta^2}{1-\eta^2} = \frac{\gamma}{1+\sqrt{1+m\gamma}}. \tag{5.14}$$

The equation for the holdup ratio is therefore

$$h = \frac{Q_o/Q_w}{V_o/V_w} = \frac{\gamma}{\frac{\gamma}{1+\sqrt{1+m\gamma}}} = 1+\sqrt{1+m\gamma}. \tag{5.15}$$

For very viscous oil, $m \approx 0$, and $h \approx 2$.

Now let us consider the holdup ratio for wavy core flow in the system of coordinates in which the wall is stationary. Given a periodic function $f(x-ct)=f(z)$ with wave length L, the average diameter of the oil core is R_1, where

$$\frac{1}{L}\int_0^L f^2(z)dz = R_1^2. \tag{5.16}$$

The input rate for a rigid core is related to the wave speed by

$$Q_o = \pi R_1^2 c = \frac{c\pi}{L}\int_0^L f^2(z)dz \tag{5.17}$$

Since both water and oil flow rates are specified, the total flow rate at each cross section is constant:

$$Q_{total} = Q_o + Q_w = \pi f^2(z)c + 2\pi\int_{f(z)}^{R_2} rv(r,z)dr \tag{5.18}$$

where $v(r,z) = \begin{cases} 0 & r = R_2 \\ c & r = f(z) \end{cases}$ and the flow inputs Q_o and Q_w are independent of z.

Then, the water input can be written as

$$Q_w = Q_{total} - Q_o = \pi c\left[f^2(z) - R_1^2\right] + 2\pi \int_{f(z)}^{R_2} rv(r,z)dr \tag{5.19}$$

independent of z.

Equation (5.19) gives an interesting z dependent decomposition of the constant water input Q_w. We may write (5.19) as

$$Q_w = \Phi(z) + \Psi(z), \tag{5.20}$$

where
$$\Phi = \pi C\left[f^2(z) - R_1^2\right] \tag{5.21}$$

and
$$\Psi(z) = 2\pi \int_{f(z)}^{R_2} rv(r,z)dr\ . \tag{5.22}$$

Suppose that a wave crest at $z=0$ just touches the pipe wall, $f(z)=R_2$. In this case, the water flow Q_w is entirely due to the forward motion of the water trapped in troughs and

$$\Psi(z) = 0, \quad \Phi(0) = \pi c[R_2^2 - R_1^2]$$

so that $\Phi(z)$ can be said to represent the trapped water. On the other hand, for perfect core flow $\Phi = 0$ for all z.

When Q_o and Q_w are given, the wave speed c and the average diameter of the oil core depends on the wave shape $f(z)$. The holdup ratio for wavy core flow is then given by

$$h = \frac{Q_o/Q_w}{R_1^2/(R_2^2 - R_1^2)} = \frac{\pi c R_1^2}{\pi c\left[f^2 - R_1^2\right] + 2\pi \int_f^{R_2} rvdr} \frac{R_2^2 - R_1^2}{R_1^2}. \tag{5.23}$$

For perfect core annular flow, the flow rate of trapped water is zero since the core radius is uniform, $f^2 = R_1^2$. Using W from (5.8) we get $h=2$. Let us focus on the flow rates at the cross section of the wave crest, where $f = f_{\max}$ and assume that $f_{\max} = R_2$. Then the integral in (5.23) vanishes and the holdup ratio is 1. Therefore, the holdup ratio for wavy core flow is between 1 and 2.

In the transition from a perfect core flow to a wavy core flow, the wave troughs will carry extra water even if the average diameter of the oil core is unchanged; this increases the water flow rate. However, when the water flow rate is fixed, the system can not increase the water flow rate. Therefore, the average diameter of the core will increase, reducing the water flow rate. In the wavy flow, more oil is in the pipe than in perfect core flow, and the holdup ratio is less than 2. Of course, the speed of the core must decrease when there is more oil and oil flow rate is fixed.

The pressure gradient β is related to the difference in area of the core at a crest and average area, this is measured by

$$d = \pi c[f^2(0) - R_1^2]. \tag{5.24}$$

The gap between the pipe wall and wave crest is smaller when d is larger, provided that the volume of oil in the pipe is fixed. Smaller gaps imply high friction and large values of the pressure gradient.

6. Computational Solution of the Wavy Core Flow

The axisymmetric core annular flow with the deformable oil core is governed by Eq. (4.11) subject to the normal stress condition specified in Eq. (4.13) and the force balance on the oil core described in Eq. (4.14). For each given value of the parameter triplet (c, R_l, h), computational solution of these equation is carried out to determine the flow of water, the shape and location of the free surface of the oil core, and the wavelength. This calculation involves an iterative solution between the calculation of the flow fieldof water and the calculation of the free surface shape. In the following discussion, important details of these two steps and the overall solution algorithm are described.

6.1. COMPUTATION OF THE FLOW FIELD OF WATER

The velocity and the pressure field involve solution of momentum and continuity equation (Eq. (4.11)) for the specified wave speed c and the available free surface shape and wavelength. Relevant details of the discretization method and the solution technique and the procedure for the determination of the pressure gradient β are now described.

6.1.1. *Discretization Method and Solution Technique*

The control volume based computational method of Patankar (1980) is used for the solution of the Navier-Stokes equations governing the flow of water. In this method, the domain of interest is divided into a set of control volumes. Values of scalar unknowns including pressure are stored at the main grid points. A staggered grid is used for storing the velocity components to avoid the occurrence of checker boarding of the pressure field. Thus, a normal velocity component is stored on each control volume face. This gives rise to momentum control volumes in the z- or r-directions to be displaced in the z- or r-directions respectively. The discretization equations for z- or r-direction velocity components are constructed by integrating the z- and r-direction momentum equations over the control volumes staggered in z- and r-directions respectively. The continuity equation is discretized over the main control volume. The convective-diffusive fluxes over the control volume faces are computed using the Power-law scheme (Patankar, 1980). The resulting discretization method expresses perfect conservation over individual control volumes and the entire domain.

Two important issues need to be addressed in the application of this discretization method for predicting the flow field of water – the representation of the free surface and the treatment of the periodicity conditions. In the present study, an axisymmetric cylindrical grid is used to discretize the entire domain ($0 \leq r \leq R_2$, $0 \leq x \leq L$). The rigid core is represented by imposing a zero velocity on the control volumes that lie in the oil core through the use of a high viscosity. This procedure approximates the wavy interface using a stepped grid. A grid independence study was carried out to determine the size of the grid necessary for accurate prediction of the water flow and the interface shape using increase number of grid points until the accuracy of the pressure distribution no significant change. The prediction of the core annular flow is carried assuming that the deformation of the wavy interface is spatially periodic. This enables prediction of flow over segment of the pipe corresponding to one wavelength. Thus, all variables in Eq. (4.11) are periodic at $z =$

0 and L. During discretization, the control volume faces at $z = 0$ and L are treated as topologically coincident to incorporate this periodicity condition (Patankar, Liu, and Sparrow, 1977).

The discretized momentum and continuity equations are solved using the SIMPLER algorithm (Patankar, 1980) that addresses the velocity-pressure coupling effectively. The algorithm involves sequential solution of the pressure, momentum, and pressure-correction equations. The line-by-line method is used for the solution of the discretization equations for each variable. The circular Tri-Diagonal-Matrix-Algorithm (TDMA) is used for solution of the discretization equations along lines in the periodic direction.

6.1.2. *Determination of the Pressure Gradient β*

Since we have chosen to specify the wave speed c, the corresponding pressure gradient β in Eq. (4.11) has to be calculated. The condition of force balance on the oil core expressed in Eq. (4.14) provides a natural method for its determination. Thus, in each iteration of the SIMPLER procedure for calculating the water flow field, the value of β is updated according to Eq. (4.14) based on the available pressure P_2 and the shear stress τ on the free surface of the oil core. At convergence of the internal iterations for the calculation of the water flow field, the value of β is determined for the specified wave speed c and the available free surface shape.

6.2. DETERMINATION OF THE FREE SURFACE SHAPE

Computational prediction of the free surface shape involves discretization and solution of the normal stress equation with an iterative adjustment of the surface shape for obtaining the prescribed average core radius R_1 and the holdup ratio h. Important details of these steps are now described.

6.2.1. *Discretization and Solution of the Normal Stress Condition*

The shape of the interface is governed by the normal stress and pressure jump condition reproduced below.

$$\frac{d^2 f}{dz^2} - \frac{1+\left(\frac{df}{dz}\right)^2}{f} + \frac{1}{\sigma}\left(1+\left(\frac{df}{dz}\right)^2\right)^{\frac{3}{2}}\left(C_p - p(z)\right) = 0 \tag{6.1}$$

The solution of this equation is sought for the available pressure variation $p(z)$ on the free surface that is determined from calculation of the water flow field. The unknown shape $f(z)$ is represented by discrete values of f at the same locations in z-direction used in the calculation of the flow field of water. The equations for these values of $f(z)$ are constructed by integrating the above equation over the main control volumes in the z-direction. The last term in the equation is treated explicitly as a source term and is assumed to be constant over the control volume. The resulting discretization equation has the following form.

$$a_i f_i = b_i f_{i+1} + c_i f_{i-1} + S\Delta z_i \tag{6.2}$$

where $$b_i = \frac{1}{z_{i+1} - z_i}, \quad c_i = \frac{1}{z_i - z_{i-1}}, \quad a_i = b_i + c_i + \frac{1 + \left(\frac{df^*_i}{dz}\right)^2}{f^{*2}_i} \Delta z_i \ ,$$

$$S_i = \frac{1}{\sigma}\left(1 + \left(\frac{df^*_i}{dz}\right)^2\right)^{\frac{3}{2}} (Cp - p_i^{**}),$$

and $$\Delta z_i = (z_{i+1} - z_{i-1})/2.$$

Similar to the flow field calculation, the periodicity of f_i values is accounted for in the above equations by recognizing that in the equation for f_N, the f_{i+1} is replace by f_1 while in the equations for f_1, the f_{i-1} is replaced by f_N. The single * in Eq. (6.2) represent available values that are updated within the inner iteration for determining the free surface shape while the **s on p_i denote that these values are kept constant during the free surface calculation and updated only in the outer iteration.

6.2.2. *Adjustment for Fixed R_1 and h*

The unknown pressure jump C_p and the wavelength L provide the two degrees of freedom necessary to determine the free surface shape consistent with the specified values of the average plug radius R_1 and the holdup ratio h. After each iteration during calculation of the f_i values, the value of C_p is increased or decreased according to whether the available f_i values imply a value of R_1 larger or smaller than that desired. Similarly, the wavelength L is increased or decreased if the current value of the holdup ratio h is larger or smaller than its prescribed value. The amount of adjustment in the values of C_p and L is determined using the secant method. It uses the predictions from the last two iterations to determine the sensitivity of R_1 and h to changes in C_p and L. The sensitivity coefficients are then used for inferring the changes in C_p and L to be made in the next iteration. At convergence, this procedure provides a free surface shape and location having the desired R_1 and h for the surface pressure variation determined from the flow field calculation.

6.3. OVERALL SOLUTION ALGORITHM

The overall solution method involves an outer iteration between the flow field calculation for water and the determination of the free surface and is outlined below.

1. Prescribe the values of wave speed c, average core radius R_1, and the holdup ratio h.
2. Assume a free surface shape. Calculate the velocity and pressure fields in the water region for the specified wave speed c. During each iteration of the flow field calculation, the pressure gradient β is adjusted to satisfy the force balance on the oil plug.
3. The shape of the free surface is determined by solving the equation describing the normal stress condition for the surface pressure determined from step 2. The wavelength and the pressure jump are adjusted in each iteration so that at

convergence the free surface shape is determined for the prescribed average core radius R_1 and holdup ratio h.

4. The new free surface is now used in determining the flow field in step 2. Thus, steps 2 and 3 are repeated till convergence to obtain a self-consistent flow field of water and free surface shape of the oil core for the prescribed values of the parameter triplet (c, R_1, h).
   The overall solution method correctly predicted the perfect core flow. Further, it predicted the same free surface shape in the flow field of water irrespective of the initial guess surface. This constituted a rigorous test for the correctness of the computational technique. Consequently, the above method was applied for computing the details of the wavy core flow for a range of the parameter triplet (c, R_1, h).

7. The Selected Waves from the Simulation and Comparison with Experiments

Our simulations are for the case in which the density of oil and water are the same; when they are not the same and the pipe is horizontal, the oil core will rise or sink. Some representative wave shapes, which look like those in experiments, are for density matched flows in Figure 7.1.

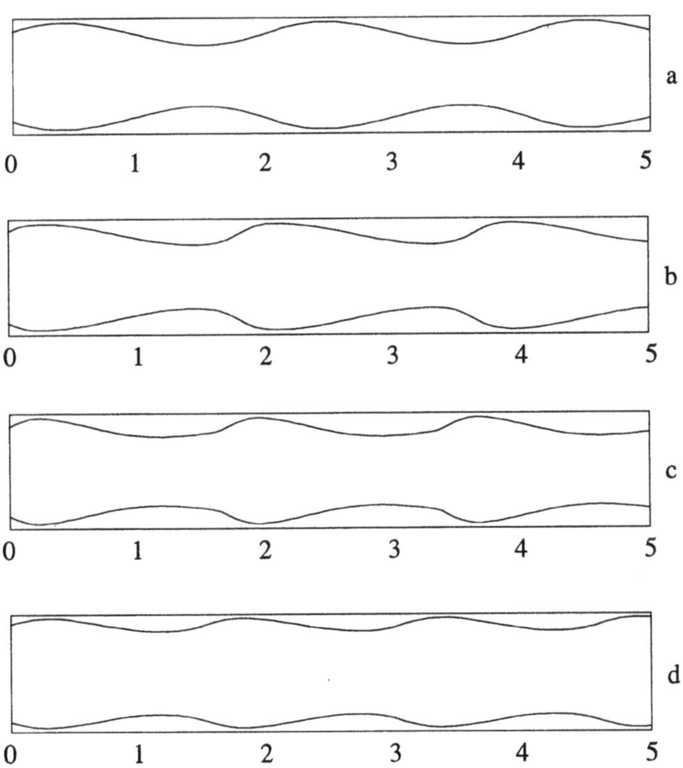

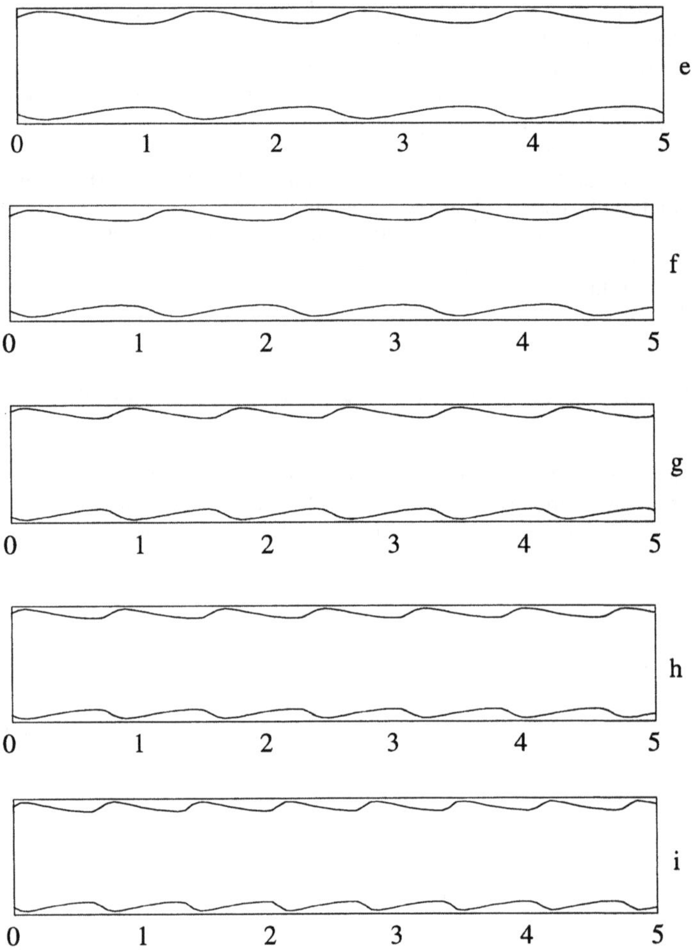

Figure 7.1. Selected wave shapes for water lubricated axisymmetric flow of oil and water with the same density ρ=1.0g/cm^3, μ_2=0.01 poise and σ=26dyne/cm for oil and water. The pipe diameter is R_2=1.0cm. Q_o and Q_w are in cm^3/sec. The data for each frame is given as the triplet (R_1, Q_o, Q_w), a (0.37, 4.30, 2.73), b (0.37, 8.6, 5.47), c (0.37, 17.2, 10.93), d (0.41, 10.56, 3.67), e (0.41, 18.48, 6.43), f (0.41, 26.41, 9.19), g (0.43, 34.85, 8.76), h (0.43, 43.57, 10.96) and i (0.43, 69.71, 17.53). The core is stationary and the wall moves to the right.

The wave shapes in Figure 7.1 are very similar the shapes in the experiment of a horizontal pipe shown in Figure 7.2. When flow rates are very high, the wave shapes are become sharper and the wavelengths became shorter. The numerical simulation exactly agrees with the prediction in Feng et al. [1995], but disagrees with the lubrication theory which was introduced by Ooms et al. [1984] and Oliemans et al. [1986].

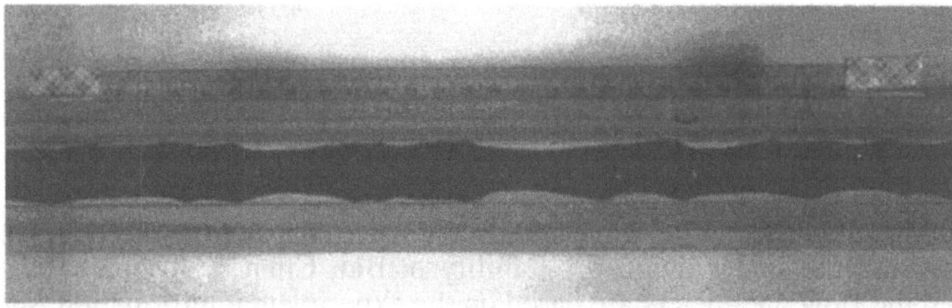

(a)

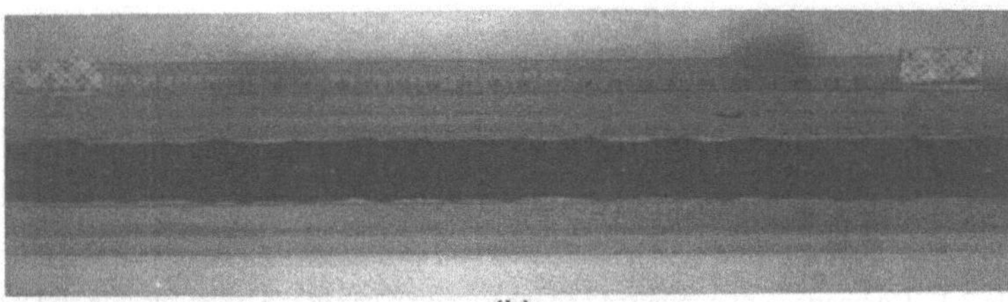

(b)

Figure 7.2. The wavy core moves from left to right. The pictures were taken in a 2.54 cm horizontal pipeline with #6 fuel oil. (a) less oil flow rate, (b) more oil flow rate at fixed water flow rate.

Bai, Chen and Joseph [1992] carried out experiments and calculated stability results for vertical axisymmetric core flow in the case when the buoyant force and pressure force on the oil are both against gravity (up flow). We observed "bamboo" waves for the oil (ρ_o=0.905g/cm^3 and μ_o=6.01 poise) in water (ρ_w=0.995g/cm^3 and μ_w=0.01 poise). We have simulated the same flow with the same parameters except that our core is infinitely viscous. The results show that μ_o=6.01 is not yet asymptotically infinitely viscous, but nevertheless the agreements are satisfactory. The equations that we used for our simulation are follow. In the water, we have

$$\begin{aligned}
&\frac{1}{r}\frac{\partial}{\partial r}(ru)+\frac{\partial w}{\partial z}=0\\
&\rho_2\left(u\frac{\partial u}{\partial r}+w\frac{\partial u}{\partial z}\right)=-\frac{\partial p}{\partial r}+\mu_2\left(\frac{1}{r}\frac{\partial}{\partial r}\left(r\frac{\partial u}{\partial r}\right)+\frac{\partial^2 u}{\partial z^2}-\frac{u}{r^2}\right)\\
&\rho_2\left(u\frac{\partial w}{\partial r}+w\frac{\partial w}{\partial z}\right)=\beta+\rho_c g-\frac{\partial p}{\partial z}-\rho_2 g+\mu_2\left(\frac{1}{r}\frac{\partial}{\partial r}\left(r\frac{\partial w}{\partial r}\right)+\frac{\partial^2 w}{\partial z^2}\right).
\end{aligned} \tag{7.1}$$

The pressure in the water is

$$P_2(r,z)=-\beta z+p(r,z)-\rho_c gz+C_2 \tag{7.2}$$

while the pressure in the core is

$$P_1=-\beta z-\rho_c gz+C_1 \tag{7.3}$$

where g is gravity and ρ_c is the composite density of the mixture

$$\rho_c = \rho_1\eta^2 + (1-\eta^2)\rho_2, \tag{7.4}$$

and $$\eta = \frac{R_1}{R_2}.$$

We compared wavelengths, wave speeds and wave shapes from our computation with experiments and the linear theory of stability in Bai, Chen & Joseph [1992]. In our comparison, the flow parameters are based on the experimental information, such as flow rates of oil and water, oil volume ratio and holdup ratio. In Bai, Chen and Joseph [1992], the holdup ratio is a constant 1.39 and the volume ratio of the oil yields the following formula:

$$\frac{R_1^2}{R_2^2} = 1 - \frac{1}{1+0.72(Q_o/Q_w)}. \tag{7.5}$$

The results are given in Table 7.1.

Table 7.1. Comparison of computed and measured values of the wave speed c and wave length L with theory of stability. In column I and II, data from Bai, Chen & Joseph [1992] were computed by linear theory of stability. In column I, the computations were carried out for fixed values of Q_o and Q_w. In column II, the calculations were done for fixed values of Q_o and volume ratio of oil.

no.	input flowrate		computations		experiments		I (PCAF$_1$)		II (PCAF$_2$)	
	Q_o	Q_w	$L(cm)$	$c(cm/s)$	$L(cm)$	$c(cm/s)$	$L(cm)$	$c(cm/s)$	$L(cm)$	$c(cm/s)$
1	25.38	13.17	1.32	55.59	1.21	57.7	0.82	79.84	0.79	52.02
2	18.19	13.17	1.66	46.45	1.31	43.28	0.92	80.21	0.96	42.54
3	11.01	13.17	1.70	37.30	1.41	35.65	1.22	79.76	1.22	33.51
4	7.42	13.17	1.33	32.73	1.22	27.81	1.65	77.00	1.33	29.42
5	7.42	6.46	1.77	20.88	1.374	19.16	1.56	58.91	1.25	17.94
6	11.01	6.46	1.66	25.45	1.79	22.90	1.23	58.12	1.16	22.17
7	14.60	6.46	1.39	30.02	1.34	28.22	1.05	54.80	1.02	26.68
8	18.19	6.46	1.15	34.59	1.17	31.06	0.95	50.85	0.87	31.33
9	21.78	6.46	0.96	39.17	0.90	36.25	0.86	49.38	0.79	35.71

The comparison of computed and measured values of the wave speed and wavelength of bamboo waves is given in Table 7.1. The computed values are slightly larger than the measured values, due to the fact that motion in the core is neglected with better agreement for faster flow.

Computed wave shapes and the observed shapes of bamboo waves are compared in Figure 7.3-7.5. The pictures were taken in a vertical pipeline with motor oil ($\rho_o = 0.905 gm/cm^3, \mu_o = 6.01 poise$) and water ($\rho_w = 0.995 gm/cm^3, \mu_w = 0.01 poise$). Both water and oil flow against gravity. The water flow rate is fixed at 200 *cm³/min* while oil flow rate is 429, 825 and 1216 *cm³/min* respectively.

The computed and observed shapes are alike. In fast flow, the velocity in the oil core is small compared with the velocity in the annulus and the oil core can be considered to be a rigid deformable body. In slower flow (Figure 7.3), the flow inside the core is not so much smaller than the flow in annulus and the stems of the waves are more readily stretched by buoyancy. Even in this case the agreements are satisfactory.

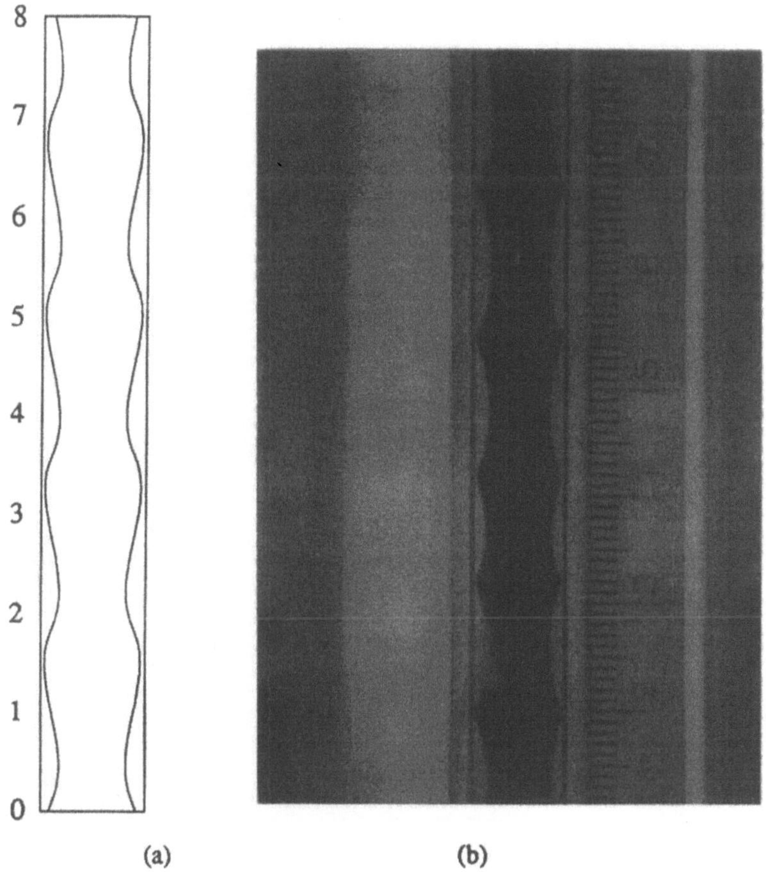

Figure 7.3. Computed waves (a) and bamboo waves (b), when $[Q_W, Q_O]$ = [200, 429] cm^3/min.

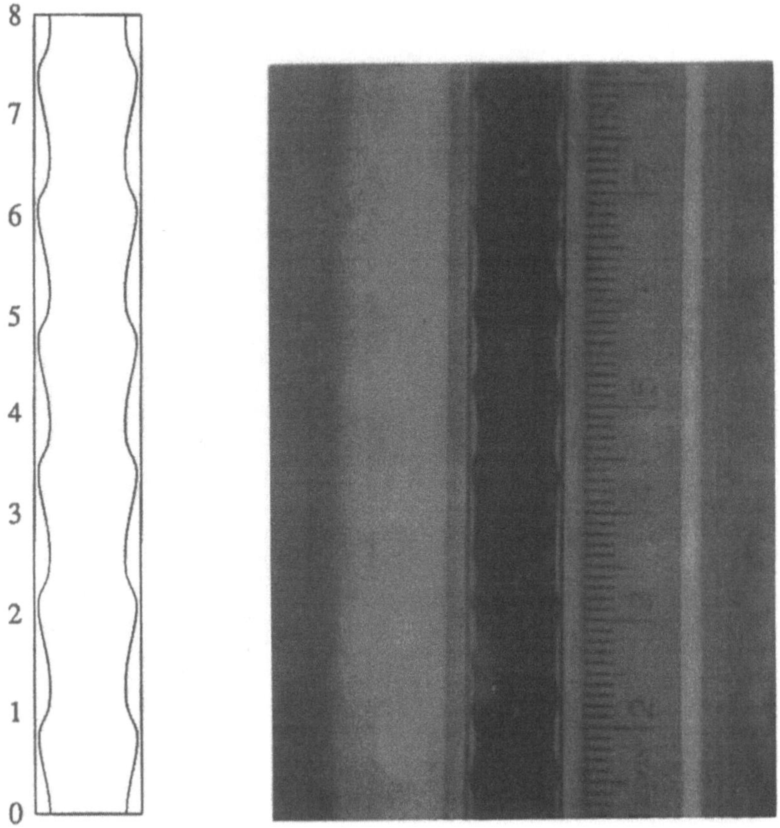

(a) (b)

Figure 7.4. Computed waves (a) and bamboo waves (b), when $[Q_W, Q_O] = [200, 825]$ cm^3/min.

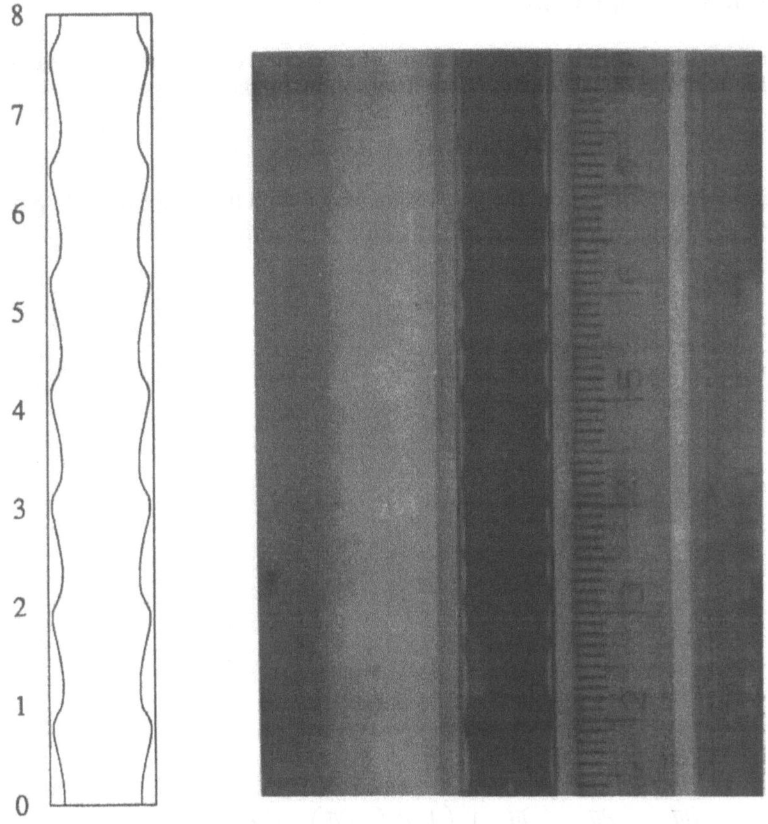

(a) (b)

Figure 7.5. Computed waves (a) and bamboo waves (b), when $[Q_w, Q_o] = [200, 1216]$ cm^3/min.

8. Dimensionless Equations

Analysis of this simulation is most useful when carried out in terms of dimensionless variables. In the dimensional equations, we used following parameters:

$$(R_2,\ R_1,\ \mu_2,\ \rho_2,\ \sigma,\ Q_o,\ Q_w\ ).$$

In the dimensionless formulation, the lengths are scaled with the pipe radius R_2, pressures are scaled by $\rho_2 U^2$, and velocities are scaled with U. Therefore

$$u = U\bar{u} \tag{8.1}$$

$$w = U\bar{w} \tag{8.2}$$

$$r, z, f, L = R_2\bar{r}, R_2\bar{z}, R_2\bar{f}, R_2\bar{L}, \tag{8.3}$$

$$R_1^2 = \frac{1}{L}\int_0^L f^2 dz = \frac{R_2^2}{\bar{L}}\int_0^{\bar{L}} \bar{f}^2 d\bar{z} \tag{8.4}$$

$$\eta^2 = \frac{R_1^2}{R_2^2} = \frac{1}{\bar{L}}\int_0^{\bar{L}} \bar{f}^2 d\bar{z}. \tag{8.5}$$

Then (4.11) becomes

$$\begin{aligned} &\frac{1}{\bar{r}}\frac{\partial}{\partial r}(\bar{r}\bar{u}) + \frac{\partial \bar{w}}{\partial \bar{z}} = 0 \\ &\bar{u}\frac{\partial \bar{u}}{\partial \bar{r}} + \bar{w}\frac{\partial \bar{u}}{\partial \bar{z}} = -\frac{\partial \bar{p}}{\partial \bar{r}} + \frac{1}{\hat{\mathbb{R}}}\left(\frac{1}{\bar{r}}\frac{\partial}{\partial \bar{r}}\left(\bar{r}\frac{\partial \bar{u}}{\partial \bar{r}}\right) + \frac{\partial^2 \bar{u}}{\partial \bar{z}^2} - \frac{\bar{u}}{\bar{r}^2}\right) \\ &\bar{u}\frac{\partial \bar{w}}{\partial \bar{r}} + \bar{w}\frac{\partial \bar{w}}{\partial \bar{z}} = \beta - \frac{\partial \bar{p}}{\partial \bar{z}} + \frac{1}{\hat{\mathbb{R}}}\left(\frac{1}{\bar{r}}\frac{\partial}{\partial \bar{r}}\left(\bar{r}\frac{\partial \bar{w}}{\partial \bar{r}}\right) + \frac{\partial^2 \bar{w}}{\partial \bar{z}^2}\right) \end{aligned} \tag{8.6}$$

where

$$\hat{\mathbb{R}} = \frac{\rho_2 R_2 U}{\mu_2}. \tag{8.7}$$

We prefer the Reynolds number

$$\mathbb{R} \overset{def}{=} \frac{\rho_2 (R_2 - R_1) c}{\mu_2}, \tag{8.8}$$

hence; the relationship between U and c is

$$c = \frac{UR_2}{(R_2 - R_1)} = \frac{U}{1-\eta}, \tag{8.9}$$

and a dimensionless wall speed is

$$\bar{c} = \frac{c}{U} = \frac{1}{1-\eta}. \tag{8.10}$$

At the boundary

$$\bar{u} = \bar{w} = 0 \qquad \text{at } \bar{r} = \bar{f}, \tag{8.11}$$

$$\bar{u} = 0, \qquad \bar{w} = \frac{1}{1-\eta} \qquad \text{at } \bar{r} = 1. \tag{8.12}$$

The normal stress balance equation becomes

$$[[\bar{P}]] = \frac{S}{\sqrt{1+\left(\frac{d\bar{f}}{d\bar{z}}\right)^2}} - \frac{S\frac{d^2\bar{f}}{d\bar{z}^2}}{\left(1+\left(\frac{d\bar{f}}{d\bar{z}}\right)\right)^{3/2}} \tag{8.13}$$

where
$$S = \frac{\sigma}{\rho_2 U^2 R_2} = \frac{J}{\mathbb{R}^2}, \tag{8.14}$$

and
$$J = \frac{\rho_2 R_2 \sigma}{\mu_2^2}. \tag{8.15}$$

The dimensionless oil flow rate

$$\bar{Q}_0 = \frac{Q_0}{UR_2^2} = \frac{\pi\bar{c}}{\bar{L}}\int_0^{\bar{L}} \bar{f}^2 d\bar{z} = \pi\bar{c}\eta^2 = \frac{\pi\eta^2}{1-\eta}, \tag{8.16}$$

is determined if η is given.

The dimensionless water flow rate may be expressed by the holdup ratio h using (5.2)

$$\bar{Q}_w = \frac{Q_w}{R_2^2 U} = \frac{1}{R_2^2 U}\int_{R_1}^{R_2} 2\pi r c_w dr = \frac{1}{h}\int_{\eta}^{1} 2\pi\bar{c}\bar{r}d\bar{r} = \frac{\pi(1-\eta^2)}{(1-\eta)h} = \frac{\pi(1+\eta)}{h}. \tag{8.17}$$

Therefore, only four parameters are required for a complete description of our problem:

$\mathbb{R}$, η, J and h .

All possible problems of scale up can be solved in this set of parameters.

9. Flow Field in Wave Troughs

When the flow is driven by pressure, there is friction between the core and wall which induces a secondary motion most easily seen in a frame moving with the core. Figure 9.1 shows these secondary motions for $\mathbb{R}$=0, $\mathbb{R}$=300 and $\mathbb{R}$=750. The flow has two parts; a more or less straight flow from left to right and an eddy. There are two points where the flow separates or rejoins the main flow. The high pressure at the front of the crest of the wave propagating into the water appears to be associated with a dividing streamline, while the low pressure at the back of the crest of the wave is related to rejoining the streamline. The wave forms are more symmetric when $\mathbb{R}$=0 and the pressure variations are moderate with positive pressure on the right at separation points and slightly larger negative pressures at the left, at reattachment points. Overall, the pressure force for $\mathbb{R}$=0 is negative and the eddy is more or less centrally located. When $\mathbb{R}$=750, the wave form, the secondary motion and the pressure distribution are profoundly influenced by inertia. The forward slope of the wave steepens and the rearward slope relaxes, the eddy moves down and to the forward face; the stagnation pressures there grow hugely while the reattachment pressures decline. All this results in a strongly positive pressure force.

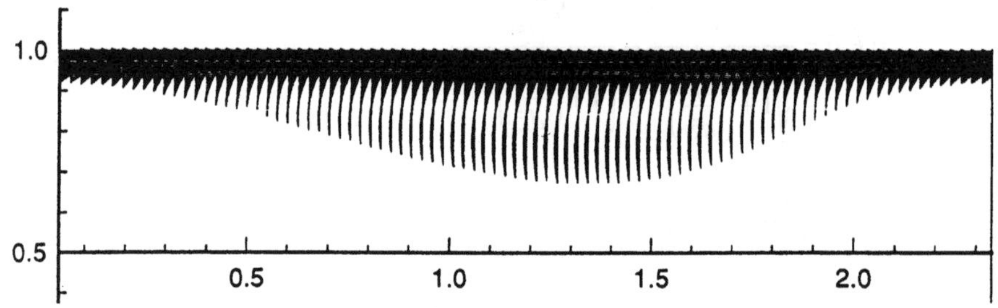

(a)

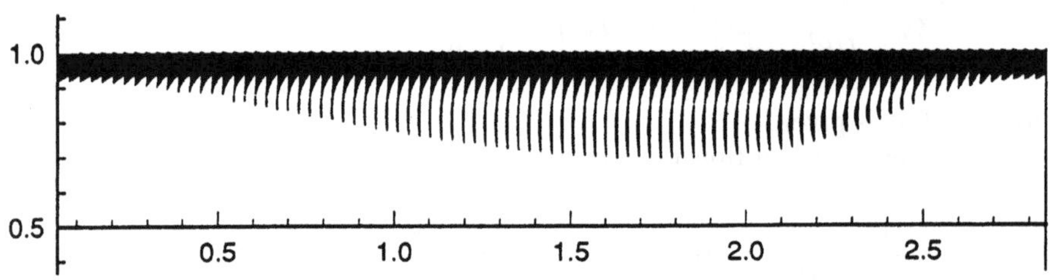

(b)

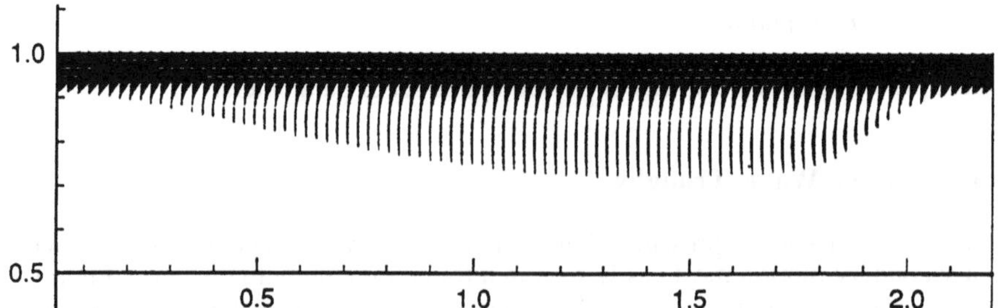

(c)

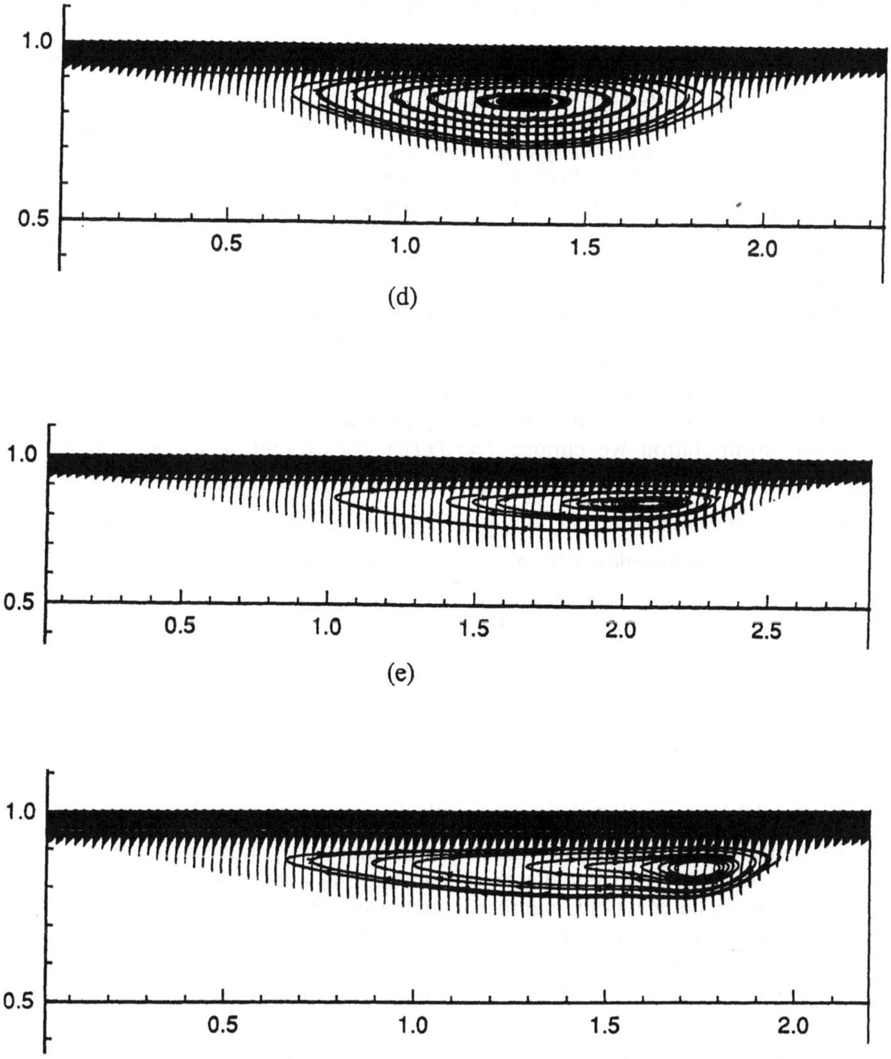

(f)

Figure 9.1. Secondary motions for (a) [$\mathbb{R}$, η, h, J] = [0, 0.8, 1.4, 13x10^4], (b) [$\mathbb{R}$, η, h, J] = [300, 0.8, 1.4, 13x10^4], (c) [$\mathbb{R}$, η, h, J] = [750, 0.8, 1.4, 13x10^4]; The pressure at the stagnation point on the steep slope at the right corresponds to the pressure maximum shown in Figure 10.2, (d) the eddies for Stokes flow as the same condition of (a), (e) the eddies as the same condition of (b), and (f) the eddies as the same condition of (c) .

10. Variation of the Flow Properties with Parameters

Now we shall show how wavelength, pressure gradients, pressure distributions on the interface and wave shape vary with $\mathbb{R}$, η, h, J.

We have already established that for a highly viscous core in which the oil moves as a rigid body, the holdup ratio varies between $h=2$ for perfect core flow and $h=1$ for the "waviest" possible core flow. In experiments, a single unique h is selected when the flow inputs are prescribed so that all but one of the family of solutions for $1<h<2$ are apparently unstable. The stable flow selects a certain $h=\tilde{h}$ and a certain wave length $\tilde{L}=L(Q_o,Q_w,\tilde{h})$. This wavelength appears to be associated to a degree with the length of wave that leads to the maximum rate of growth of small disturbances perturbing perfect core annular flow. The holdup ratio for the bamboo waves which appear in up flow in the vertical pipeline studied in the experiments of §7 was about 1.39 independent of the inputs Q_w and Q_o and the same $h=1.39$ is attained in down flow at large Reynolds numbers. These observations have motivated us to compute many results for $h=1.4$.

In our computation we choose $J=13x10^4$ corresponding to the actual physical parameters in wavy core flow in water in which μ_2=0.01 poise ρ=1 g/cm^3, σ=26 dyne/cm, in a pipe of a $1cm$ diameter. A value of η=0.8 is fairly typical of experiments. The definition of the dimensionless parameter p^* and gradient β^* are

$$p^*=\frac{p(f(z),z)-P_2(f(0),0)}{\mu^2}\rho R_2^2, \tag{10.1}$$

and

$$\beta^*=\frac{\rho R_2^2}{\mu^2}\beta. \tag{10.2}$$

Figure 10.1 shows that the wavelength decrease with $\mathbb{R}$ and the pressure gradient increase linearly with $\mathbb{R}$ for fixed representation values of η, h and J.

Figures 10.2 and 10.3 show that the wave steepens at the front and relaxes at the back of a wave crest as the Reynolds number is increased. The steepening is produced by the high pressures at the stagnation point on the wave front (of Figure 9.1). The pressure force, the area under the pressure curve, changes sign for $10<\mathbb{R}<150$. The variation of the pressure force and pressure peak with $\mathbb{R}$ is very nearly linear. We expected inertia to lead to a variation with $\mathbb{R}^2$ but evidently the high pressures that are generated by inertia are dissipated by deforming the interface, at least when $\mathbb{R}\leq 750$.

Figures 10.4, 10.5 and 10.6 shown how the wavelength, the pressure gradient, pressure distribution and wave forms vary with h when [η, $\mathbb{R}$, J] = [0.8, 600, 13x10^4]. The wave shape is more unsymmetric and the pressure force is greater when h is close to 1. The wavelength is a decreasing of h when $\mathbb{R}$=600, but is an increasing function of h when $\mathbb{R}$=0 for the same parameter (cf. Figure 10.12).

Figures 10.7, 10.8 and 10.9 shown how the wavelength, the pressure gradient and wave forms vary with η when [$\mathbb{R}$, h, J] = [600, 1.4, 13x10^4]. The wavelength decreases and the positive pressure peak and wave front slope all increase as the gap get smaller. This suggests that the levitating pressure force will intensify as the gap get smaller when the density of the oil and water are different.

Figures 10.10, 10.11 and 10.12 show how the wavelength, the pressure gradient, pressure distribution and wave forms vary with h when $[\eta, \mathbb{R}, J] = [0.8, 0, 13x10^4]$. The wave shape is more unsymmetric and the pressure force is greater when h is close to 1. The wavelength is an increasing function of h when $\mathbb{R}=0$.

Figures 10.13, 10.14 and 10.15 show how the wavelength, the pressure gradient and wave forms vary with η parameter when $[\mathbb{R}, h, J] = [0, 1.4, 13x10^4]$. The pressure force is negative under all conditions, and it is even more negative when gap is small. The wave forms nevertheless are steeper on the front than on the rear face, though this asymmetry is less pronounced than at higher Reynolds numbers.

Figure 10.16 shows that the wavelength increases with surface tension, which smoothes the wave.

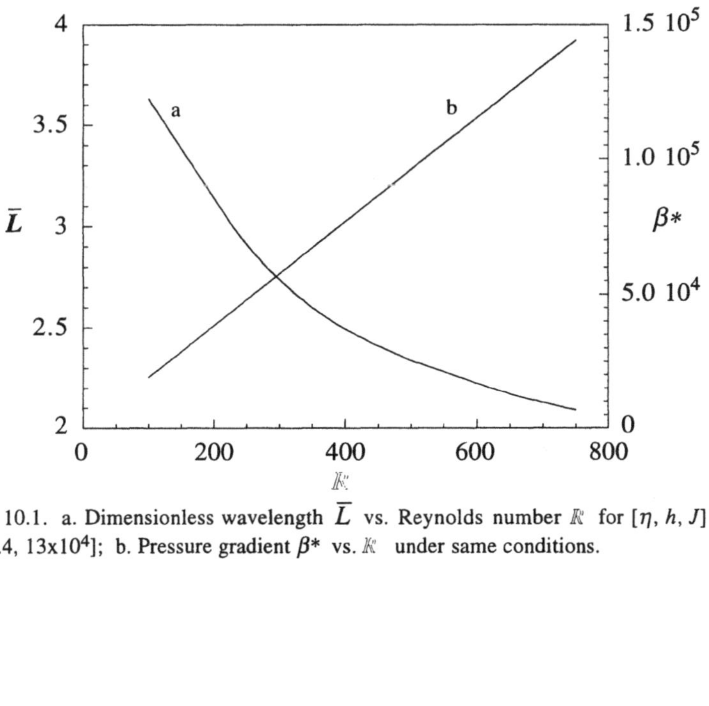

Figure 10.1. a. Dimensionless wavelength $\bar{L}$ vs. Reynolds number $\mathbb{R}$ for $[\eta, h, J] = [0.8, 1.4, 13x10^4]$; b. Pressure gradient $\beta*$ vs. $\mathbb{R}$ under same conditions.

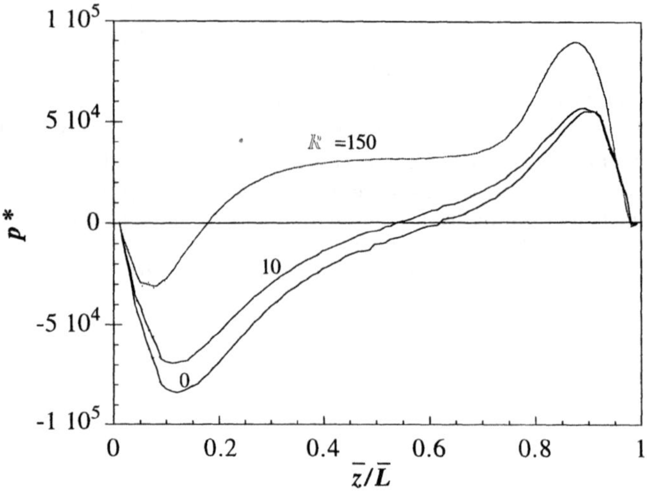

(a)

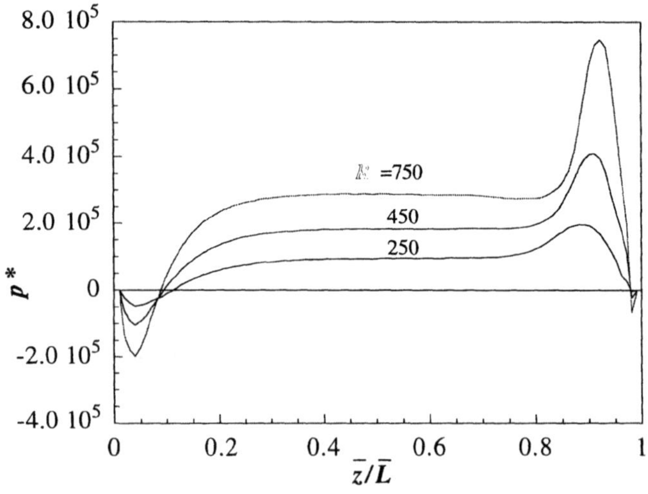

(b)

Figure 10.2. (a) Pressure distributions on the interface $p^*(\bar{z}/\bar{L})$ for $\mathbb{K}$ =0, 10, 150 when [η, h, J] = [0.8, 1.4, 13x10^4]. Note that the pressure force, the area under the pressure curve, is negative for $\mathbb{K}$ =0, 10 and is positive when $\mathbb{K}$ =150. (b) Pressure distributions on the interface for $\mathbb{K}$ =250, 450, 750 when [η, h, J] = [0.8, 1.4, 13x10^4]. All the pressure forces are positive with the greatest pressure at the forward points of stagnation.

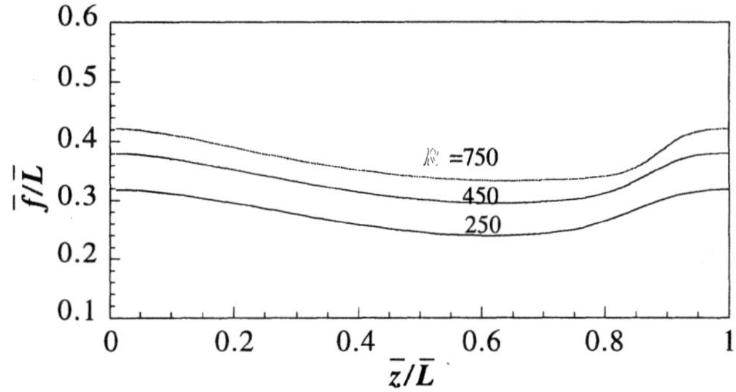

Figure 10.3. Wave shape ($\bar{f}$ / $\bar{L}$ vs. $\bar{z}$ / $\bar{L}$) for three ℝ =250, 450, 750 when [η, h, J] = [0.8, 1.4, 13x10^4].

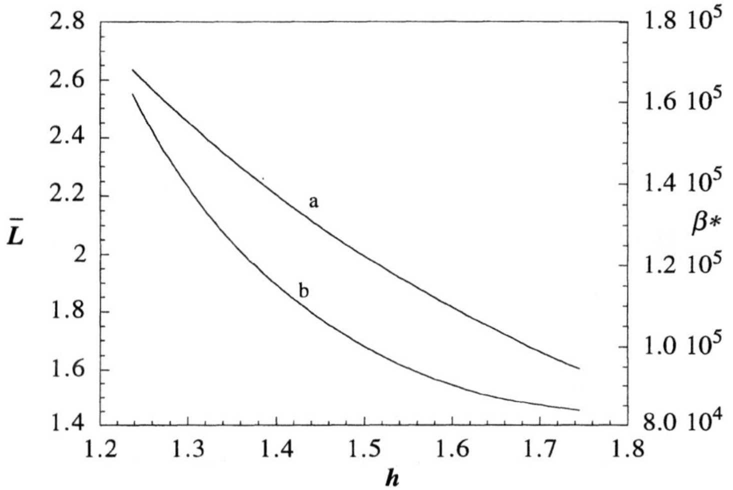

Figure 10.4. a. Wavelength $\bar{L}$ vs. h for [η, ℝ , J] = [0.8, 600, 13x10^4]; b. Pressure gradient β* vs. h under the same conditions.

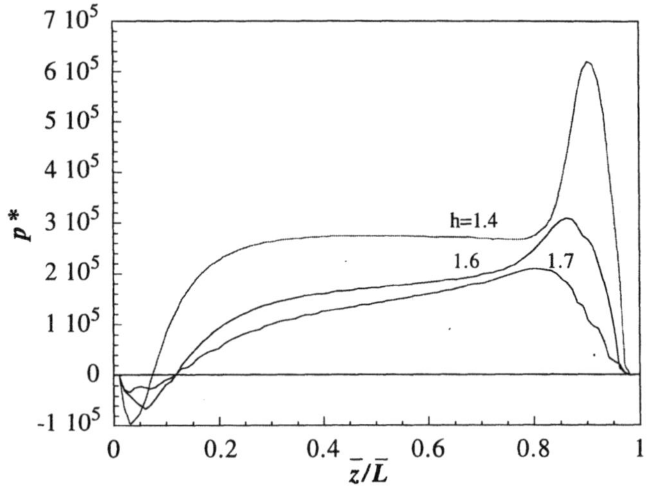

Figure 10.5. Pressure distributions on the interface for different h when $[\eta, \mathbb{K}, J] = [0.8, 600, 13\text{x}10^4]$.

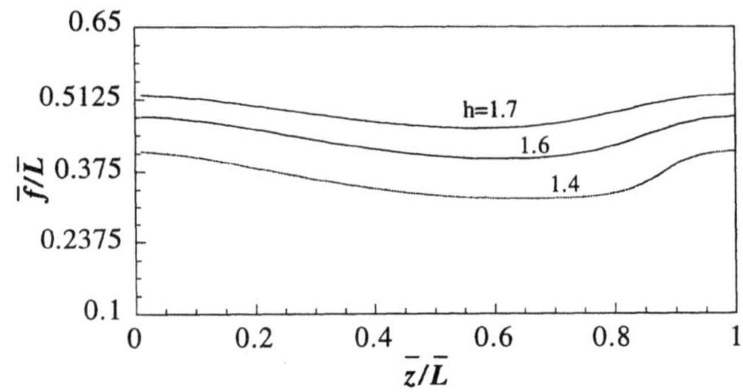

Figure 10.6. Wave shape $\bar{f}\,/\,\bar{L}$ vs. $\bar{z}\,/\,\bar{L}$ under the conditions specified in Figure 10.5.

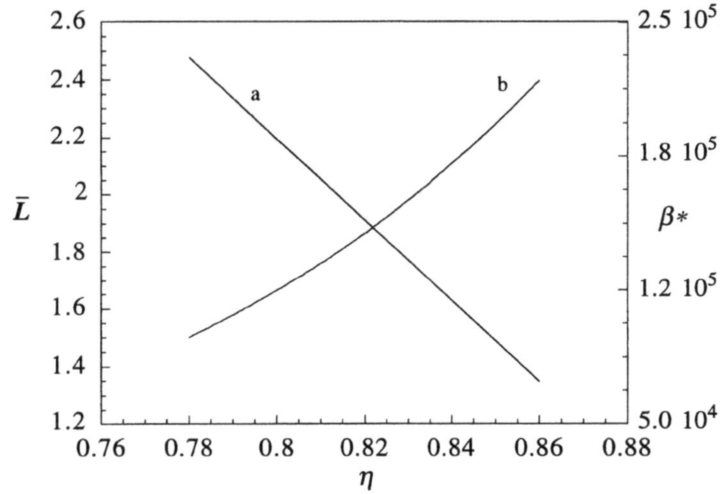

Figure 10.7. a. Wavelength $\bar{L}$ vs. η for [$\mathbb{R}$, h, J] = [600, 1.4, 13x10^4]; b. Pressure gradient β* vs. η under the same conditions.

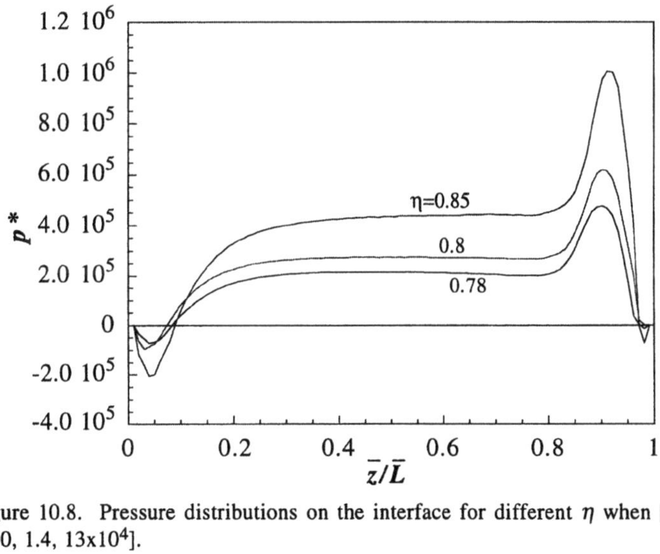

Figure 10.8. Pressure distributions on the interface for different η when [$\mathbb{R}$, h, J] = [600, 1.4, 13x10^4].

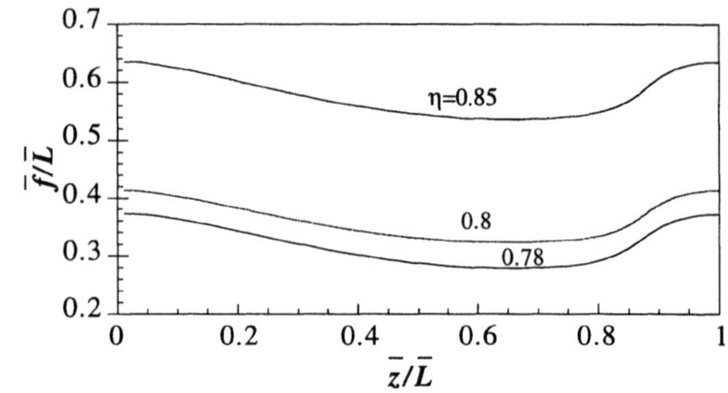

Figure 10.9. Wave shapes for the conditions of Figure 10.8.

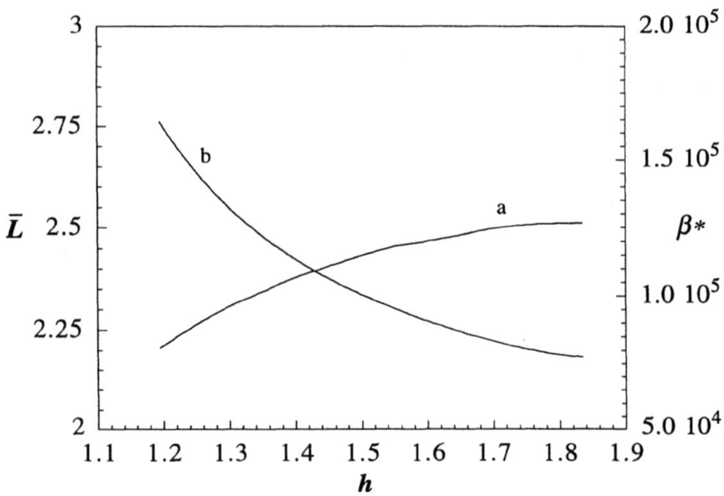

Figure 10.10. a. Wavelength $\bar{L}$ vs. h for $[\eta, \mathbb{R}, J] = [0.8, 0, 13\times10^4]$; b. Pressure gradient β^* vs. h under the same conditions.

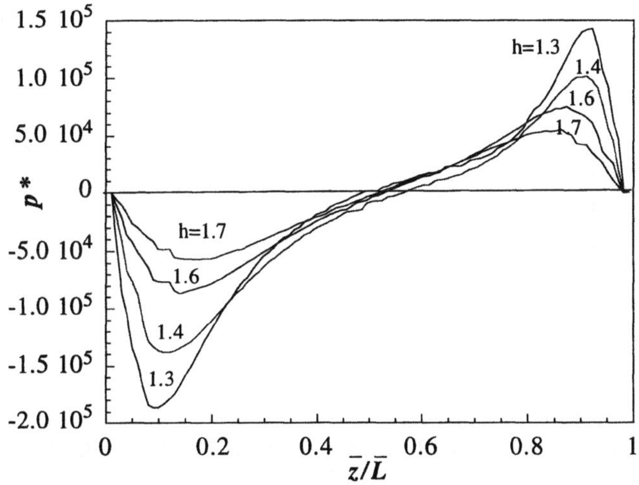

Figure 10.11. Pressure distributions on the interface for different h when [η, $\mathbb{R}$, J] = [0.8, 0, 13x10^4].

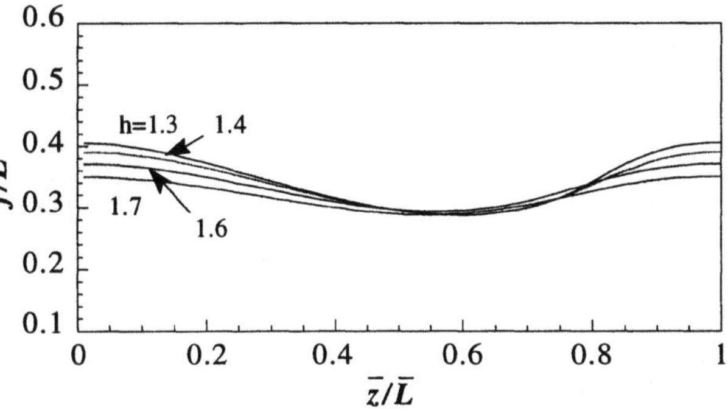

Figure 10.12 Wave shape $\bar{f} / \bar{L}$ vs. $\bar{z} / \bar{L}$ under the conditions specified in Figure 10.11.

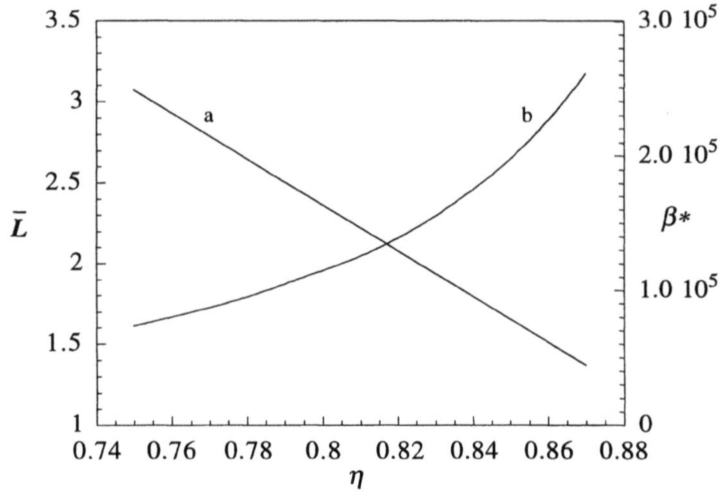

Figure 10.13. a. Wavelength $\overline{L}$ vs. η for $[\mathbb{K}, h, J] = [0, 1.4, 13\mathrm{x}10^4]$; b. Pressure gradient β^* vs. η under the same conditions.

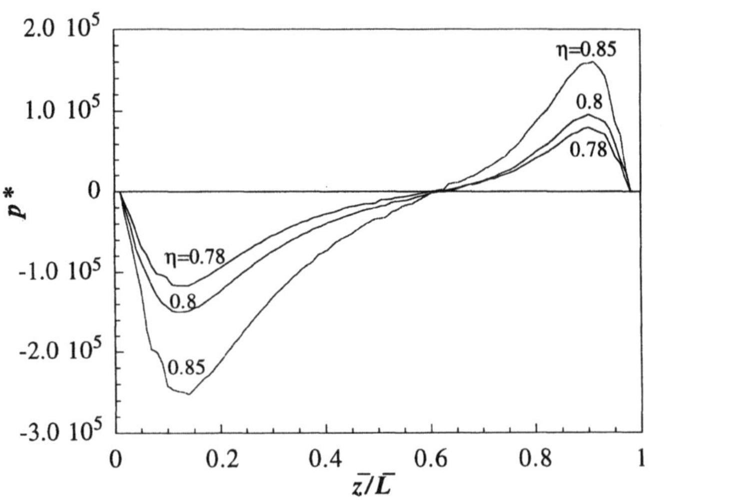

Figure 10.14. Pressure distributions on the interface for different η when $[\mathbb{K}, h, J] = [0, 1.4, 13\mathrm{x}10^4]$.

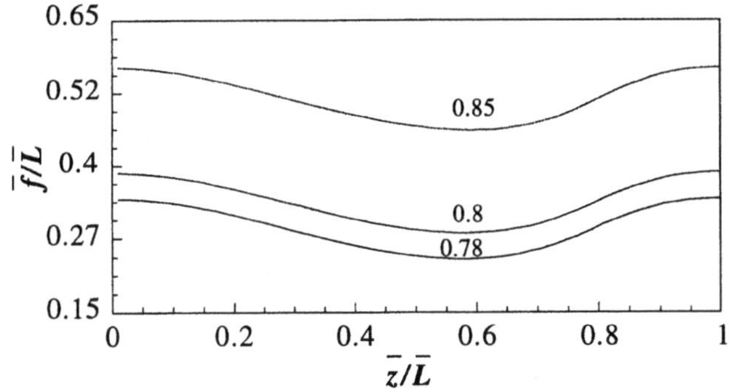

Figure 10.15. Wave shapes for the conditions of Figure 10.14.

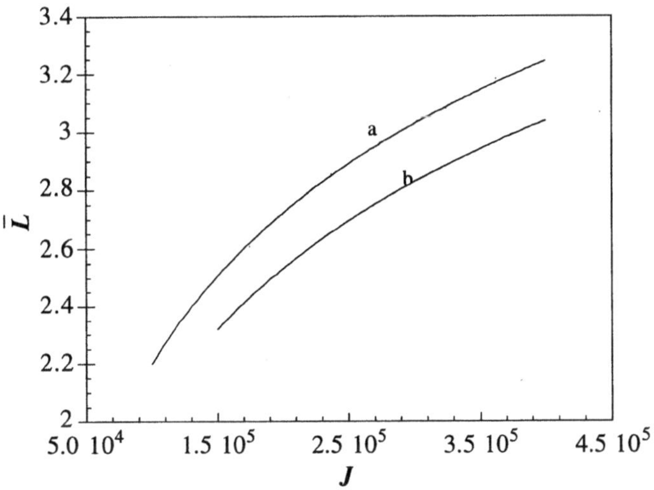

Figure 10.16. Wavelength $\bar{L}$ vs. J a $[\mathbb{K}, \eta, h] = [0, 0.8, 1.4]$; b $[\mathbb{R}, \eta, h] = [600, 0.8, 1.4]$.

11. Threshold Reynolds Numbers and Levitation of Wavy Core Flows

The total pressure force on the core is an integral of the pressure on the core surface, the area under the pressure curves shown, say, in Figures 10.2, 10.5, 10.8, 10.11 and 10.14. The total pressure force is negative at zero and small Reynolds numbers, and is more and more positive as the Reynolds number is increased past a threshold. Figure 11.1 shows that the wavelength $\overline{L}(h)$ is an increasing function of h when $\mathbb{R}=0$ and is a decreasing function of h when $\mathbb{R}$ is greater than some threshold.

The concept of a threshold may be formulated in terms of the pressure force

$$F_r = \frac{2\pi}{\overline{L}} \int_0^{S_{\overline{L}}} p^* \boldsymbol{n} \bullet \boldsymbol{e}_r \frac{ds}{dz} dz \tag{11.1}$$

where $\boldsymbol{n}$ is the normal to the core S is the arc length. Figure 11.2 gives the locus of points on where $F_r = 0$, with $F_r < 0$ below and $F_r > 0$ above. The threshold value is always positive though it decreases with h, since $F_r < 0$ for Stokes flow. In Figure 11.3 we plotted F_r for h=1.9 where η=0.9. This is a case in which lubrication theory might be used; however, $\mathbb{R}_c$ is about 2.25 so that the pressure force is negative for lubrication theory (see §12). Figure 11.4 shows that the wavelength changes with hold-up ratio for [η, $\mathbb{R}$, J] = [0.8, 100, 13x10^4]. The slope dL/dh is greater than zero in a lower hold-up ratio and less than zero in a higher hold-up ratio.

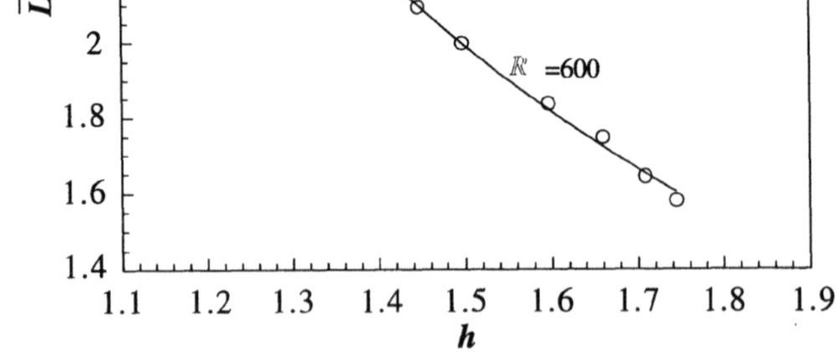

Figure 11.1. Wavelength $\overline{L}$ vs. h for a [η, $\mathbb{R}$, J] = [0.8, 0, 13x10^4], and for b [η, $\mathbb{R}$, J] = [0.8, 600, 13x10^4]

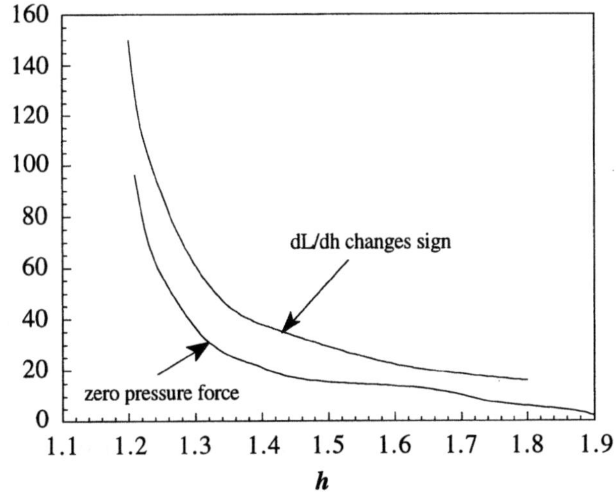

Figure 11.2. Threshold Reynolds number $\mathbb{R}$ vs. h for $[\eta, J] = [0.8, 13\text{x}10^4]$.

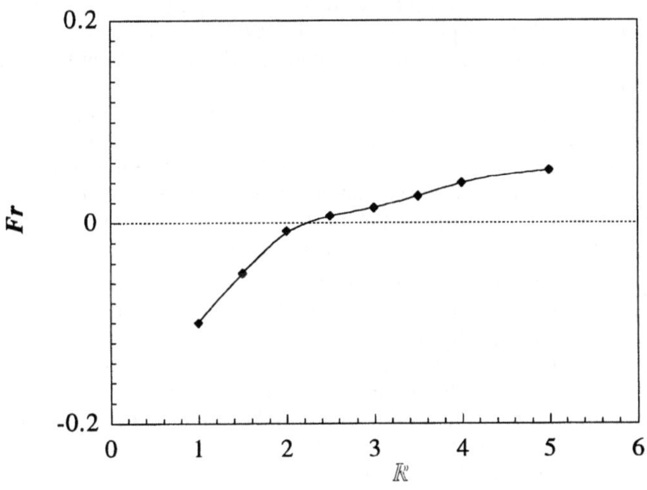

Figure 11.3. Pressure force due to pressure on the interface vs. $\mathbb{R}$ for $[\eta, h, J] = [0.9, 1.9, 13\text{x}10^4]$.

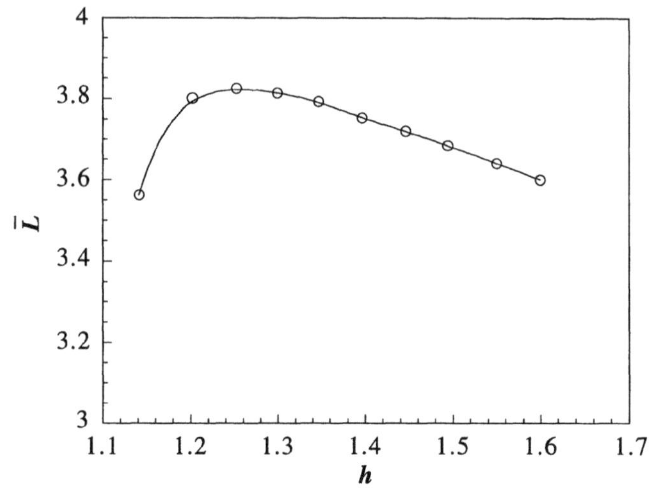

Figure 11.4. The slope *dL/dh* of the wavelength changes with holdup ratio for [η, $\mathbb{R}$, $\mathbb{J}$]=[0.8, 100, 13x10^4].

We may argue that a positive levitation of a lighter or a heavier than water core flow requires a positive pressure force to push the core away from wall. The pressure forces are associated with the form of waves that they generate. Waves also develop in Stokes flow, but the pressure forces associated with these waves are negative like those on the reversed slipper bearing which pull the slipper to the wall. When the Reynolds number is higher than the threshold, high positive pressures are generated, especially at the stagnation point on the steep part of the wave front (Figure 9.1).

In the axisymmetric problem with matched density considered here, lateral motions of the core off center are not generated by pressure forces, whether they be positive or negative, because the same pressure acts all around the core. We may consider what might happen if the core moved to a slightly eccentric position due to a small difference in density. The pressure distribution in the narrow part of gap would intensify and the pressure in the wide part of the gap would relax according to the predicted variation of pressure with η shown in Figure 10.8. In this case a more positive pressure would be generated in the narrow gap which would levitate the core. The equilibrium position of the core would then be determined by a balance buoyancy and levitation by pressure force which, in the case of matched densities, would center the core. In eccentric horizontal core flows of lighter-than-water oils which are in experiments, the waves in the small gap near the wall are shorter and the positive pressures are higher than in the large gap at the bottom, the speed of the core is increased, the core will move to the center and the shape of the wave will tend to the axisymmetric ones studied here.

12. Lubrication Theory

Lubrication theory is valid when inertia is neglected (Stokes flow), when the wave amplitude is small and the radial velocity u and $\partial w/\partial z$ are negligible. The last conditions imply that secondary motions are not present or are very weak. The required conditions can be achieved in the limit in which $\mathbb{R}$ and $2\text{-}h$ tend to zero, in Stokes flows which perturb perfect core flows. Small gaps are one way to achieve small $\mathbb{R}$, but other possibilities are compatible with lubrication theory. It is of value to examine the lubrication theory in bright light since it is very popular with applied mathematicians and has played an historically important role in the development of the theory of core flows.

After applying the assumptions of lubrication theory, the governing equations (4.11) reduce to

$$0=\beta-\frac{\partial p}{\partial z}+\mu_2\frac{1}{r}\frac{\partial}{\partial r}\left(r\frac{\partial w}{\partial r}\right). \tag{12.1}$$

where $w=0$ at $r=f(z)$ and $w=-c$ at $r=R_2$. Hence

$$w(r)=\frac{1}{4\mu}\frac{dP_2}{dz}(r^2-f^2(z))-\frac{c+\dfrac{1}{4\mu}\dfrac{dP_2}{dz}(R^2-f^2(z))}{\ln\left(\dfrac{R_2}{f}\right)}\ln\left(\frac{r}{f}\right), \tag{12.2}$$

where $$P_2=-\beta z+p(z)+P_2(0), \tag{12.3}$$

and $$\frac{P_2(0)-P_2(L)}{L}=\beta. \tag{12.4}$$

The pressure difference is opposed by the integral of the shear stress on the wall,

$$\beta L\bullet\pi R_2^2=2\pi R_2\int_0^L\mu_2\left.\frac{dw(r)}{dr}\right|_{r=R_2}dx \tag{12.5}$$

After combining (12.2) and (12.5), we relate the wave speed c to the pressure gradient

$$c=-\frac{1}{4\mu_2}\frac{dP}{dz}(R_2^2-f^2). \tag{12.6}$$

This equation can be also obtained from (5.5) with $\mu_1\to\infty$. For a given speed c, the pressure gradient is determined by the shape of the core.

$$\frac{dP_2}{dz}=-\beta+\frac{dp}{dz}=-\frac{4\mu c}{R_2^2-f^2(z)} \quad . \tag{12.7}$$

Since p is a periodic function, the driving pressure gradient is given by

$$\beta = \frac{P_2(0) - P_2(L)}{L} = \frac{1}{L}\int_0^L \frac{4\mu c}{R_2^2 - f^2(z)} dz. \tag{12.8}$$

The periodic pressure, a functional of *f(z)*, is

$$p(z) - P_2(0) = \beta z - \int_0^z \frac{4\mu c}{R_2^2 - f^2(z)} dz. \tag{12.9}$$

In the perfect core annular flow $f(z)=R_1$ is uniform and (12.8) and (12.9) show that $p(z)-P_2(0)=0$ and pressure gradient will be constant..

We carried out an analysis of these equation by assuming *f(z)* and computing *p(z)*. Of course, the normal stress balance is not satisfied by the assumed shape but we could iterate all of the assumed shapes to unique which satisfies the reduced normal stress balance (4.13)

$$P_1 - P_2 = \frac{\sigma}{f\sqrt{1+f'^2}} - \frac{\sigma f''}{(1+f'^2)^{3/2}} \quad . \tag{12.10}$$

Here

$$P_1 = -\beta z + P_1(0) \tag{12.11}$$

in the core and (12.7) implies that

$$P_2 = -\int_0^x \frac{4\mu c}{R_2^2 - f^2(z)} dz - P_2(0) \tag{12.12}$$

in the annulus. The pressure jump across the interface is then

$$P_1 - P_2 = -\beta z + \int_0^z \frac{4\mu_2 c}{R_2^2 - f^2(z)} dz + C_p, \tag{12.13}$$

where $\quad C_p = P_1(0) - P_2(0)$.

Our iteration starts with any one of the three trial waves, say $f_1(z)$. Then we compute

$$P_1 - P_2 = -\beta z + \int_0^z \frac{4\mu_2 c}{R_2^2 - f_1^2(z)} dz + C_p \tag{12.14}$$

and carry out the first iteration using the normal stress balance to compute $f_2(z)$:

$$-\beta z + \int_0^z \frac{4\mu_2 c}{R_2^2 - f_1^2(z)} dz + C_p = \frac{\sigma}{f_2\sqrt{1+f_2'^2}} - \frac{\sigma f_2''}{(1+f_2'^2)^{3/2}} \quad . \tag{12.15}$$

We then compute

$$P_1 - P_2 = -\beta z + \int_0^z \frac{4\mu_2 c}{R_2^2 - f_2^{\,2}(z)} dz + C_p$$

and so on. This iteration converges to the unique solution shown in Figure 12.1 for each of three very different guesses for $f_1(z)$.

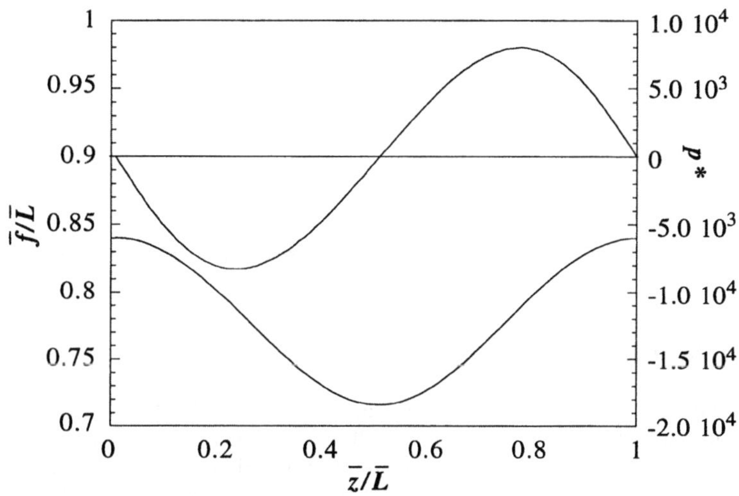

Figure 12.1. Pressure distribution and wave shape from lubrication theory when $[\eta, h, J] = [0.8, 1.9, 13\text{x}10^4]$.

The wave is nearly symmetric, but the pressure force is slightly negative. Figure 12.2 compares the pressure distribution from lubrication theory and Stokes flow; in both cases the area under the pressure curve is negative. To use lubrication theory as a predictive tool, it is necessary at least to study the stability of the flow with h near 2, in most cases they will be unstable. Stokes flows may be unstable to off-center perturbations of the type considered by Huang and Joseph [1994] at small but finite Reynolds numbers since the unbalanced core could possibly be sucked to the wall by the negative pressure force.

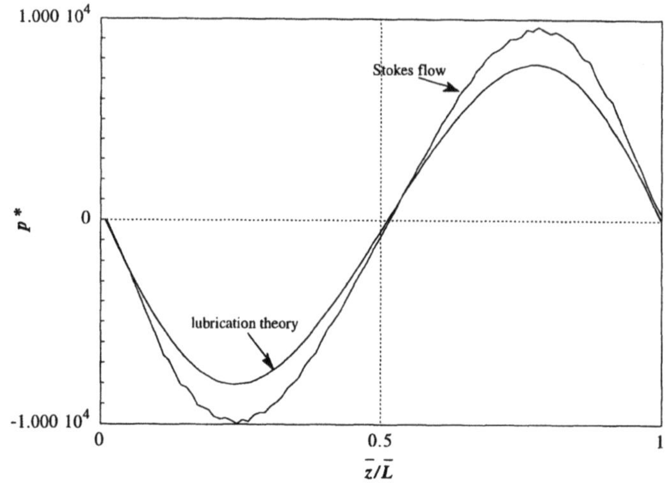

Figure 12.2. Comparison of pressure distributions when [η, h, J]= [0.8, 1.9, 13x10^4].

13. Conclusions

Core annular flows of liquids with the same density and a high viscosity ratio were computed in a direct numerical simulation. It was assumed that flow is axisymmetric and the core is solid with adveceted periodic standing interfacial waves fixed on the moving core. These assumptions reduce the number of parameters defining the problem to four : Reynolds number, radius ratio, holdup ratio and surface tension parameter. In dimensional terms, for given material parameters, we get solutions when the volume flow rates of oil, water and the holdup ratio are prescribed. Only the flow rates are given in experiments and the holdup is then determined by stability, so we are computing a family of the solutions most of which are unstable.

The numerical solutions have the properties predicted by Joseph in Feng et al.; the high pressures on the forward facing slope of the wave, where the water enters into a converging gap, steepen the wave and reduction of the pressure on the lee side smooths the wave. This leads to unsymmetric waves unlike those which levitate a slipper bearing. The problem of levitation does not arise in the density matched core, but the pressure distributions which actually develop in this case seem to be such as to center a slightly displaced core only when the Reynolds number is greater than a threshold value which depends on the parameters but in all cases is strictly positive. The concept of a threshold Reynolds number for levitation is suggested but not established by our work. To establish such a concept it would be sufficient to see if solutions for the density matched computed here are stable to off-center perturbations and to compute and consider the stability of finite amplitude off-center core flows when the density is not matched.

The important parameters of the waves usually includes wave speed, frequency, wavelength and wave number. In the computation, the wave speed was specified. If the wavelength is obtained, all parameters can be determined. Therefore, we only focus on the wavelength. There are four major parameters which influence the wavelength, and they are (Reynolds number), J (dimensionless interfacial tension), h (holdup ratio) and η (radius ratio). The higher Reynolds number causes larger pressure differences at the stagnation points and strong secondary motions in the wave trough, greater interfacial tension induces a longer wavelength, the holdup ratio changes the pressure distribution, and the larger volume of core, the larger friction between wave core and the pipe wall.

If we keep interfacial tension, volume ratio and holdup ratio as constants, the higher Reynolds number will induce a higher pressure difference between the oil and water. When the Reynolds number increases and the pressure difference is very large, the interfacial tension cannot balance the pressure. The wavelength will be decreased, which increases the curvature of the interface in order to keep the force balanced. The shapes of the wave is also different from variation of the Reynolds number. A higher Reynolds number also causes a unsymmetric wave shape while the Stokes flow have a nearly sysmmetric wave shape. The front side of the wave is sharper and the wavelength is shorter at the higher Reynolds number, while the wave is more flat and the wavelength is longer at lower Reynolds numbers. The dimensionless overall pressure gradient will decrease with Reynolds number increase. However, the dimensional pressure increases with the Reynolds number increase.

The interfacial tension plays an important role in the wave flow. At a certain flow condition, the interface is easily deformed if the interfacial tension is small. A greater interfacial tension requires greater pressure difference to deform the interface. Therefore, a smaller interfacial tension has a shorter wavelength compared to a larger interfacial tension because the shorter wavelength will have larger curvature. Generally speaking, larger interfacial tension tends to make the oil core smooth. There appears to be no benefit in producing a lifting force for large interfacial tension if the densities of water and oil do not match. However, large interfacial tension is good for a high flow rate situation and a large pipeline because the interface is not easy to break up.

The wavelength and pressure gradient also are functions of the radius ratio. Radius ratio determines the average gap size between oil core and water. The smaller volume ratio induces a larger gap size which causes weaker secondary motion in the wave trough and longer wavelength. A greater volume ratio results in a shorter wavelength and larger value of pressure gradient because it gives rise to more friction between the wall and core and induce stronger secondary motion.

A threshold Reynolds number is found; the pressure force of the water on the core relative to a fixed reference pressure is negative for Reynolds number below the threshold and is positive above. The wavelength increases with the holdup ratio when the Reynolds number is smaller than a second threshold and decreases for larger Reynolds numbers. The pressure gradient always decreases with holdup ratio increasing.

The streamlines between Stokes flow and inertia flow are quiet different. In the Stokes flow, the streamlines are more symmetric while the streamlines are much unsymmetric in inertia flow.

Acknowledgments

This work was supported by the Department of Energy, Office of Basic Energy Science; the National Science Foundation; and the Minnesota Supercomputer Institute. The authors wish to thank Dr. J. Feng, Adam Huang, Peter Huang, Terrence Liao and Mr. Walter Wang for their help and helpful discussions.

Reference

Bai, R., Chen, K., & Joseph, D. D.(1992)Lubricated pipelining: Stability of core-annular flow. Part 5: Experiments and comparison with theory. *J. Fluid Mech.* **240**, 97-142.

Feng, J., Huang, P. Y., & Joseph, D. D.(1995)Dynamic simulation of the motion of capsules in pipelines. *J. Fluid Mech.* **286**, 201-227.

Huang, A. & Joseph, D. D. (1995) Stability of eccentric core annular flow. *J. Fluid Mech.* **282**, 233-245.

Joseph, D. D., Nguyen, K. & Beavers, G. S.(1984) Non-uniqueness and stability of the configuration of flow of immiscible fluids with different viscosities. *J. Fluid Mech.* **141**, 319-345.

Joseph, D. D., & Renardy, Y. Y.(1993)Fundamentals of two-fluid dynamics. Springer-Verlag New York.

Liu, H.(1982) A theory of capsule lift-off in pipeline. *J. Pipelines* **2**, 23-33.

Oliemans, R. V. A.(1986) The lubricating film model for core-annular flow. Delft University Press.

Oliemans, R. V. A. & Ooms, G.(1986)Core-annular flow of oil and water through a pipeline. *Multiphase Science and Technology*. vol. **2**, eds. Hewitt, G. F., Delhaye, J. M. & Zuber, N., Hemisphere Publishing Corporation.

Ooms, G., Segal, A., Van der Wees, A. J., Meerhoff, R. & Oliemans, R. V. A.(1984) A theoretical model for core-annular flow of a very viscous oil core and a water annulus through a horizontal pipe. *Int. J. Multiphase Flow* **10**, 41-60.

Ooms, G., Segal, A., Cheung, S. Y. & Oliemans, R. V. A.(1985)Propagation of long waves of finite amplitude at the interface of two viscous fluids. *Int. J. Multiphase Flow* **11**, 481-502.

Patankar, S. V. (1980)*Numerical Heat Transfer and Fluid Flow.* Hemisphere.

Patankar, S. V., Liu, C. H., & Sparrow, E. M.(1977)Fully developed flow and heat transfer in ducts having streamwise-periodic variations of cross-section area. *J. Heat transfer*, **99**, 180.

A NON-AXISYMMETRIC PATTERN SELECTION PROBLEM IN CORE-ANNULAR FLOW

YURIKO Y. RENARDY
Department of Mathematics and
Interdisciplinary Center for Applied Mathematics
Virginia Polytechnic Institute and State University
Blacksburg, VA 24061-0123, U. S. A.

Abstract. Core-annular flow of two fluids is examined at the onset of the azimuthal mode number ±1 non-axisymmetric instability. The patterns that emerge are traveling waves and standing waves. The former travel in the azimuthal direction as well as the axial direction and would be observed as corkscrew waves. The standing waves travel in the axial direction but not in the azimuthal direction and appear as snakes. Weakly nonlinear interactions are studied to see whether one of these waves will be stable to small amplitude perturbations. Sample situations for down-flow are discussed. The corkscrews tend to be preferred when the annulus is narrow, while snakes are more likely when the annulus is wide.

1. Introduction

In perfect core-annular flow of two fluids, the core fluid has a cylindrical interface of uniform radius centered on the pipe axis, surrounded by an annulus of the outer fluid. The response of the flow to perturbations of the form $\exp(in\theta + i\alpha x + \sigma t)$, where θ denotes the azimuthal angle, x the vertical axis, t the time, is examined. When perfect core-annular flow loses stability to an axisymmetric $n = 0$ disturbance, then the interface can deform to a varicose shape (Preziosi, Chen & Joseph 1989; Joseph & Renardy 1993). On the other hand, when the system loses stability for the $n = \pm 1$ non-axisymmetric modes, the critical modes are proportional to $\exp(i\theta+i\alpha x+i\omega t)$, $\exp(-i\theta+i\alpha x+i\omega t)$, and their complex conjugates. Their mode interaction can give rise to either waves traveling in the azimuthal direction (corkscrew waves) or standing waves (snake waves). Both of these

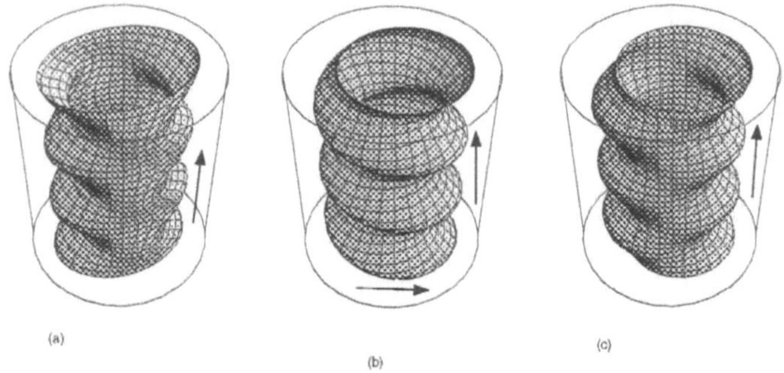

Figure 1. (a) The $n = 0$ axisymmetric mode. The $n = \pm 1$ non-axisymmetric onset can give rise to (b) corkscrews or (c) snake waves.

waves travel in the axial direction. The snake waves are simply translated along (Figure 1). This article is based on the paper of Renardy (1997).

Finite amplitude corkscrew waves have been observed in the core-annular flow of viscous crude oil and water (see photos in Chap. VII of Joseph & Renardy 1993; and Joseph *et al.* 1997). The experiments of Bai *et al.*(1992) are conducted in a vertical inverted loop. Different flow regimes are shown in the review article of Joseph *et al.*(1997). It is possible for the snake mode to be set up initially if the initial condition favors it, for instance, when there are bamboo waves in the up-flow part of the loop, so that at the top of the loop there is an azimuthal symmetry that favors the solution to be a standing wave. However, if the corkscrews are stable and the snakes not, then this flow would later evolve into the corkscrews.

Renardy (1997) addresses the question of whether corkscrew or snake waves bifurcate from perfect core-annular flows. This is a Hopf bifurcation with O(2) symmetry, i.e., the problem is invariant under reflection and translation in the azimuthal coordinate θ. In this case, it is known Ruelle (1973) that if both standing and traveling wave solutions are supercritical, then one of them is stable, and that if one is subcritical, then both are unstable. The equations governing the problem and the weakly nonlinear analysis are detailed in Appendices A and B of Renardy (1997). Section 4 outlines the application of a center manifold reduction scheme on the governing equations to derive the weakly nonlinear amplitude evolution equations for the interaction of the two critical Hopf modes. This yields two coupled Landau equations. We note however that for conditions slightly above criticality, an interval of wavenumbers becomes unstable; consequently, there will be development of the disturbances in space as well as in time, and a theory accounting for modulations would be of interest. The usual Ginzburg-Landau equation does not apply to this problem due to the presence of a long-wave mode for which the decay rate tends to zero in the limit of infinite wavelength Renardy & Renardy (1993). For core-annular flow, we would have four amplitude factors. Two of these correspond to amplitudes of the azimuthal traveling waves, the third to a long-wave modulation of the interface height, and the fourth results from the pressure. We have chosen for simplicity to begin with the Stuart-Landau framework. Throughout the analysis, the pressure gradient in the flow direction is assumed to be fixed. An alternative approach would be to keep the flow rate fixed. However, the work of Renardy (1989) on channel flow indicates that there would not be a significant difference with either formulation unless the Reynolds number is high.

In § 4, the stability criteria for the solutions are deduced. In § 5, the pattern selection analysis is carried out for sample conditions in down-flow, in which the annular fluid is the more dense. In the first example, the annular fluid is the less viscous, and the preferred pattern is found to be the corkscrews for narrow annuli and snakes for wide annuli. In the sample case with equal viscosity and free fall, there is a preference for corkscrews, unless the annulus is very wide, in which case snakes are preferred.

2. Formulation of Equations

Two immiscible liquids are flowing through a vertical pipe of radius R_2. The governing equations are the Navier-Stokes equations and incompressibility in each fluid. At the interface, the velocity and shear stresses are continuous, the jump in the normal stress is balanced by the effect of interfacial tension, and the kinematic free surface condition holds. The interface between the two liquids is given by $r^* = R(\theta, x, t)$ where (r^*, θ, x^*) denotes cylindrical coordinates, x^* points in the same direction as the centerline base velocity (we non-dimensionalize the flow so that the dimensionless centerline velocity is 1). For upward flow, gravity is acting in the negative x^*-direction, and for downward flow, it is acting in the positive x^*-direction. The flow is driven by a pressure gradient $dP^*/dx^* = -f^*$. Asterisks denote dimensional quantities. The pipe axis is at $r^* = 0$. The average value of R is denoted by R_1. The core region $0 \leq r^* \leq R(\theta, x, t)$ is occupied by fluid 1 with viscosity μ_1 and density ρ_1. Fluid 2, with viscosity μ_2 and density ρ_2, is located in the annulus $R(\theta, x, t) \leq r \leq R_2$. The following are four dimensionless parameters:

$$m = \mu_2/\mu_1, \quad a = R_2/R_1, \qquad \zeta = \rho_2/\rho_1, \quad K = (f^* + \rho_1 g)/(f^* + \rho_2 g), \tag{1}$$

where K measures the ratio of driving forces in the core and annulus.

The dimensionless variables are $(r, x) = (r^*, x^*)/R_1$, $(u, v, w) = (u^*, v^*, w^*)/W_0^*(0)$, $t = t^* W_0^*(0)/R_1$, $p = p^*/[\rho_1 W_0^{*2}(0)]$, where the centerline velocity is

$$W_0^*(0) = (\rho_2 g + f^*)\frac{R_1^2}{4\mu_2}A, \quad A = mK + a^2 - 1 + 2(K-1)\log a. \tag{2}$$

The dimensionless base velocity field is $(0, 0, W(r))$ where

$$W(r) = \begin{cases} [a^2 - r^2 - 2(K-1)\log(r/a)]/A & 1 \le r \le a \text{ (annulus)} \\ 1 - mr^2 K/A & r < 1 \text{ (core)} \end{cases} \tag{3}$$

The base pressure field satisfies

$$dP/dx = -f, \quad [\![P]\!] = J/\mathbb{R}_1^2. \tag{4}$$

The interfacial tension parameter is

$$J = T^* R_1 \rho_1/\mu_1^2, \tag{5}$$

where T^* denotes interfacial tension. This parameter is related to a surface tension parameter introduced by Chandrasekhar [1961] in his study of capillary instability of jets of viscous liquid in air. The Reynolds numbers $\mathbb{R}_i$ are defined by $\mathbb{R}_i = \rho_i W_0^*(0) R_1/\mu_i, i = 1, 2$, where $\mathbb{R}_1/\mathbb{R}_2 = m/\zeta$.

We now perturb perfect core-annular flow so that the total solution is $(u, v, W + w), P + p$, $R/R_1 = 1 + \delta(\theta, x, t)$ and consider the equations for (u, v, w, p, δ). We denote the set of unknowns (u_1, v_1, w_1, p_1, u_2, v_2, w_2, p_2, δ) by Φ. The governing equations to cubic order are written in the schematic form

$$L\Phi = N_2(\Phi, \Phi) + N_3(\Phi, \Phi, \Phi). \tag{6}$$

This is not the amplitude equation but represents the raw governing equations shown in Appendix A of Renardy (1997). The operator L incorporates the linear terms in the differential equations and bounday conditions, while N_2 stands for the quadratic terms, and N_3 the cubics. The cubic terms arise from the Taylor expansions of the interface conditions. For small amplitudes, both the quadratic and cubic terms contribute to the cubic nonlinearity in the amplitude equations to be derived in § 4. We can write the real linear operator L in the form $A + B\frac{d}{dt}$. We also introduce the notation $L(\sigma) = A + \sigma B$. The problem is characterized by six dimensionless parameters: m, a, ζ, J, K and $\mathbb{R}_1$.

3. Linear Stability Analysis

The consideration of non-axisymmetric waves in core-annular flow is mentioned in the linear stability analyses of Preziosi *et al.*(1989) and Boomkamp & Miesen (1992). Hu & Patankar (1995) have performed a linear stability analysis and found parameter regimes for unstable situations, where mode 1 has larger growth rates than mode 0 and correlated the wavelength of maximum growth rate with the observed wavelength for corkscrew waves in the experiments of Bai *et al.*(1992). In their regimes, there are bands of unstable wavelengths for modes 0 and or 1. For example, their Figure 2 shows the $n = 1$ mode to be unstable over wavenumbers less than approximately 2, and the $n = 0$ mode unstable for long waves and for wavenumbers approximately between 0.5 and 2.5, while the maximum growth rate is achieved by the $n = 1$ mode at wavenumber approximately 0.5. In this section, we show that there are critical situations where mode 1 onsets at a finite wavelength while other modes are stable. This complements the correlation with the experiments in that the n=1 mode is actually shown to be the first mode to become unstable as a bifurcation parameter is increased.

We replace $[u, v, w, p](r, \theta, x, t)$ and $\delta(\theta, x, t)$ with amplitude functions $[u, v, w, p](r)$ and an amplitude constant δ times $\exp[\imath n\theta + i\alpha\ x + \sigma t]$ in the usual way. The linearized problem is of the form $(A + \sigma B)\Phi = 0$ where the operators A and B are defined in Appendix A of Renardy (1997). The Chebyshev-tau scheme Orszag (1971) is used to discretize this system in the radial direction (Orszag (1971), Gottlieb & Orszag (1977)). The numerical results were checked against those of Hu & Patankar (1995) which include the effects of all the parameters, and Joseph, Renardy & Renardy (1983) which treats the case of zero surface tension and equal densities. Our code is independent of the latter code. Asymptotic formulas for long and short waves are contained in § VI.1 (h) and § VI.1(i), respectively, of Joseph & Renardy (1993).

4. Analysis of Pattern Selection

Bifurcation from the base solution (3) - (4) will be investigated for onset conditions. There is an eigenvalue $i\omega$ at axial wavenumber α and azimuthal wavenumber $n = 1$ and another eigenvalue $i\omega$ at the same α but azimuthal wavenumber $n = -1$. In the x and θ directions, the solution is periodic. The operator $L(\sigma)$ has a one-dimensional nullspace at the zeroth Fourier mode $(d/dx = 0, d/d\theta = 0)$ with $p =$ constant and the other variables equal to zero due to the fact that we can add an arbitrary constant to the pressure. To make the pressure unique we normalize it.

We reduce the partial differential equations and boundary conditions governing the problem to two complex-valued ordinary differential equations by applying the center manifold reduction scheme (cf. Renardy 1989, 1997) We view one of the fluid parameters, e.g., the Reynolds number, as a bifurcation parameter. For instance, we set $\lambda = Re - Re_c$, where Re_c is the critical value of a Reynolds number. Close to criticality ($\lambda = 0$) we denote by $\omega_1(\lambda)$ the eigenvalue which arises from perturbing $i\omega$, $\zeta_1(\lambda)$ the eigenfunction belonging to $\omega_1(\lambda)$ at azimuthal wavenumber $n = 1$, and by $\zeta_2(\lambda)$ the eigenfunction belonging to $n = -1$. The eigenfunction ζ_2 is retrieved from ζ_1 by the azimuthal reflection $n \to -n, v \to -v$. The eigenfunction computed in § 3 is used for $\zeta_1(0)$. The two eigenfunctions have the forms

$$\zeta_1 = \tilde{\zeta}_1(r)\exp(i\alpha x + i\theta), \quad \zeta_2 = \tilde{\zeta}_2(r)\exp(i\alpha x - i\theta). \tag{7}$$

The inner product $(\cdot,\cdot)$ at the discretized level corresponds to a weighted L_2 inner product, in which the components of the first function are complex conjugated. The eigenfunction b_1 to the adjoint problem satisfies $(b_1, L(\omega_1(\lambda))\Phi) = 0$, where Φ is arbitrary, and has the same dependence on θ and x as the respective eigenfunctions ζ_1. The adjoint eigenvector corresponding to ζ_1 has the form $b_1 = \tilde{b}_1(r)\exp(i\alpha x + i\theta)$. In weakly nonlinear theory, the solution is expanded as $\Phi \sim \sum_{i=1}^2 z_i\zeta_i + \sum_{i=1}^2 \bar{z}_i\bar{\zeta}_i + \ldots$, where z_i are complex time-dependent amplitude functions. Appendix B of Renardy (1997) contains the derivation of the amplitude equations:

$$\begin{aligned} \frac{dz_1}{dt} &= \omega_1(\lambda)z_1 + \beta_1(\lambda)|z_1|^2 z_1 + \beta_2(\lambda)|z_2|^2 z_1, \\ \frac{dz_2}{dt} &= \omega_1(\lambda)z_2 + \beta_1(\lambda)|z_2|^2 z_2 + \beta_2(\lambda)|z_1|^2 z_2, \end{aligned} \tag{8}$$

$\omega_1(\lambda) = i\omega + \epsilon_1$, where ω and the bifurcation parameter ϵ_1 are real. The equilibrium solutions are the following.

4.1. TRAVELING WAVES (CORKSCREWS)

For the traveling waves, one of the z_i is zero. The corkscrews travel in the azimuthal as well as the axial direction. We denote $z_1 = r_1\exp(ir_2t)$, where

$$r_1^2 = -\epsilon_1/\mathrm{Re}\,\beta_1, \quad r_2 = \omega + \mathrm{Im}\,\beta_1 r_1^2, \quad z_2 = 0. \tag{9}$$

The solution is supercritical if $-\epsilon_1/\mathrm{Re}\,\beta_1 > 0$, and since the base flow is stable for $\epsilon_1 < 0$, we require $\mathrm{Re}\,\beta_1 < 0$ for the solution to exist above threshold. If this condition is satisfied, then the solution branch is 'supercritical'. Since our bifurcation problem involves a two-fold degeneracy, supercriticality alone is not enough to imply the stability of that solution branch to small perturbations, and there is a further condition to be satisfied. In Renardy (1997), it is shown that for stability we require

$$\mathrm{Re}\,\beta_1 < 0, \text{ and } \mathrm{Re}\,\beta_2 < \mathrm{Re}\,\beta_1. \tag{10}$$

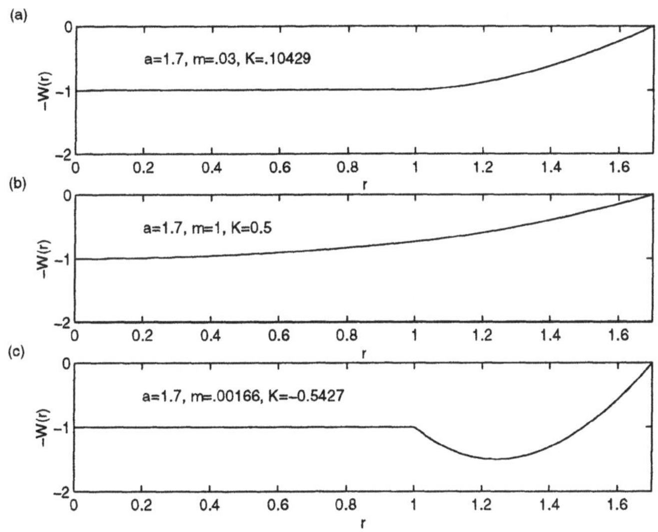

Figure 2. Base velocity profiles representing the down-flows discussed in § 5. (a) $m = 0.03, a = 1.7, K = 0.10429, \zeta = 1.1$, (b) $m = 1, \zeta = 2, a = 1.7, K = 0.5$, (c) $m = .00166, \zeta = 1.1, a = 1.7, K = -0.5427$.

4.2. STANDING WAVES (SNAKES)

The snake waves consist of both amplitudes $|z_1| = |z_2|$ being nonzero. We denote $z_1 = r_1 \exp(ir_2t)$ where

$$r_1^2 = -\epsilon_1/(\text{Re } \beta_1 + \text{Re } \beta_2), \quad r_2 = \omega + (\text{Im } \beta_1 + \text{Im } \beta_2)r_1^2. \tag{11}$$

This solution exists for $\epsilon_1 > 0$ if $(\text{Re } \beta_1 + \text{Re } \beta_2) < 0$, and this is the condition for the snake solution to be a supercritical branch. In Renardy (1997), it is shown that for stability, we require the two conditions

$$\text{Re } (\beta_1 + \beta_2) < 0, \text{ and Re } \beta_1 < \text{Re } \beta_2. \tag{12}$$

5. Results

We present results for down-flows. The base velocity profiles for the cases we consider are shown in Figures 2 (a-c). We define down-flows to be those in which gravity is acting in the same direction as the dimensional centerline velocity.

Some of the pattern selection results will be given in tabular form. At the tabulated values of the wavenumbers α, the corresponding Im σ are given. The ϵ_1 of the previous section denotes a perturbation to Reσ as the bifurcation point is traversed. The calculation of the Landau coefficients yields the preferred pattern.

5.1. CRITICAL CONDITIONS

We begin with the parameters of Fig. 3(b) of Hu and Patankar (1995), $a = 1.4$, $m = 0.01$, $\zeta = 1.1$, $\mathbb{R}_1 = 1, J = 0$. Here, the $n = 0, 1$ modes are stable for long waves, the $n = 1$ achieves the maximum growth rate at wavenumber $\alpha \approx 1.7$, and the other modes are also unstable. The effect of adding surface tension is the stabilization of the higher modes, shorter waves, and a destabilization of the axisymmetric mode due to capillary instability. We balance these effects and choose $J = 0.01$, for which the axisymmetric mode is only slightly destabilized, while the $n \geq 2$ modes are stabilized, leaving only the $n = 0, 1$ modes unstable. By varying the density ratio ζ from 0.8 to 1.1, the magnitudes of the instability changes but the signs are not affected; we choose $\zeta = 1.1$. There is more sensitivity to the radius ratio a and the viscosity ratio m. When $a = 1.5, 1.6$, mode 1 develops long-wave instability. Mode 1 is further destabilized as a increases as reported by Hu and Patankar (1995). Mode 1 is stable for $a = 1.1, 1.2, 1.3$. Slight increases in m stabilize both modes; namely, at $m = 0.02$, the window of instability for mode 0 shrinks, while that for mode 1 remains similar. At $m = 0.03$, mode 0 is stable while mode 1 retains a window of instability. At $m = 0.04$, mode 1 develops a long-wave instability as well as a separate window of O(1) wave instability $0.3 \leq \alpha \leq 1.6$. In the search for a mode 1 criticality, we retain $m = 0.03$ and examine a between 1.3 and 1.4. The delicate dependence on a is shown in Figure 3(a). In this manner a criticality is found for $a = 1.37, \alpha = 1.3, \sigma = -0.0004 - 1.26i$ (there is no round-off error in these values; in order to get $\mathrm{Re}\sigma$ closer to 0, we need to evaluate the critical values of a and α to more significant figures than we have given. These values for criticalities are sufficiently accurate for the application of the ensuing nonlinear analysis). The growth rates are shown in Figure 3(b) for modes 0 to 2. Higher modes and shorter waves are effectively stabilized by surface tension. The wavenumbers in the range $1 < \alpha < 1.5$ are all close to criticality. The Reynolds number is slightly increased in Figure 3(c), resulting in the wavenumbers around 2.5 for mode 0 coming close to criticality, and the wavenumbers 1 to 2.5 for mode 1 being slightly above onset.

The critical situation of Fig. 3(b) is close to a resonant 2:1 mode interaction; namely, mode 1 is at criticality with wavenumber $\alpha_c = 1.3$ and eigenvalue $i\omega = -1.26i$, and mode 0 is critical with wavenumber $2\alpha_c$ and eigenvalue $2i\omega$. The Reynolds number is increased slightly for Figures 3(c)-(d) to show that an interval of wavenumbers is close to onset. Figures 3(c)-

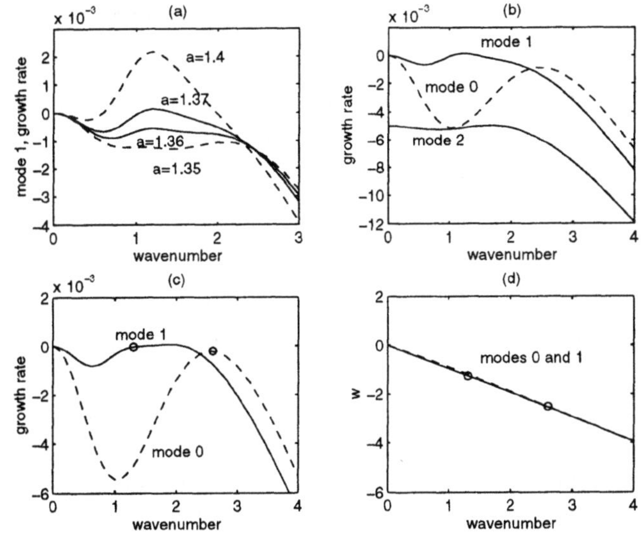

Figure 3. (a) Growth rate Re σ versus wavenumber α for $J = .01$, $\mathbb{R}_1 = 1$, $m = .03$, $\zeta = 1.1$, mode 1. Effect of varying a. (b) Growth rate versus wavenumber for $J = .01$, $m = .03$, $a = 1.37$, $\mathbb{R}_1 = 1$, $\zeta = 1.1$, $K = .10429$. Mode 1 is at criticality. Modes 0,2, and higher are stable. (c) Growth rate versus wavenumber for $J = .01$, $m = .03$, $a = 1.37$, $\mathbb{R}_1 = 1.06$, $\zeta = 1.1$, $K = .10429$. Mode 1 at $\alpha = 1.3$ and mode 0 at $\alpha = 2.6$ are resonant and circled. (d) For the parameters of (c), the imaginary part of the eigenvalue $Im\ \sigma = w$ is plotted against α. Both modes 0 and 1 coincide, and circles denote the resonant values corresponding to those of (c).

(d) show the real and imaginary parts of the eigenvalues at these wavenumbers. For the $\alpha = 1.3$, mode 1, onset, for instance, a proper analysis should incorporate the resonant interaction, and would show the existence of new solutions that involve both modes 0 and 1. A diagnostic for the occurrence of resonant situations is that the Landau coefficients become sensitive to small changes. In our case, this is due to a near-singular matrix in the calculation of the interaction term ψ_{12} (see Appendix B of Renardy (1997)), and a pole for the Landau coefficient β_2. Table 1 shows that the Landau coefficient β_1, which determines the supercriticality of the corkscrews, varies smoothly, while the imaginary part of β_2 changes from negative and large at $\alpha = 1.2$ to positive and large at $\alpha = 1.3$, indicating that there is a pole here. The last column denotes the solution type: in addition to the (c) and (s), we have situations where only one solution is supercritical and therefore both solutions are unstable. Thus, (cx) in the last column denotes the supercriticality of corkscrews but not snakes. The numerical results indicate that at resonance, there is a change from a situation where snakes are subcritical and corkscrews are supercritical to a situation where both

α	Im σ	β_1	β_2	pattern
1.0	-0.97	-12+62i	5-109i	s
1.1	-1.06	-10+59i	7-196i	s
1.2	-1.16	-8+56i	8-629i	cx
1.3	-1.26	-6+53i	-24+680i	c
1.4	-1.36	-5 +50i	-3+242i	s
1.5	-1.45	-3+49i	.5+157i	s
1.9	-1.85	-1+47i	3+87i	cx
2.0	-1.94	-1+48i	3+83i	cx

TABLE 1. $a = 1.37, \mathbb{R}_1 = 1.0$, $m = 0.03$, $\zeta = 1.1$, $K = 0.10429387$, $J = 0.01$. The last column indicates stable snakes (s) or stable corkscrews (c) when both solutions are supercritical. (cx) denotes the supercritical corkscrews when snakes are subcritical and both solutions are unstable.

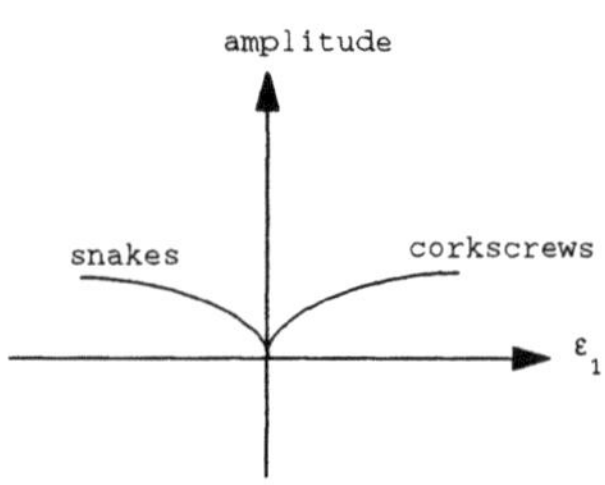

Figure 4. Bifurcation diagram close to resonance. $m = 0.03$, $\zeta = 1.1$, $J = 0.01$, $K = 0.10429387$, $a = 1.37$, $\mathbb{R}_1 = 1$, $\alpha = 1.3$, mode 1 onset.

are supercritical and corkscrews are stable. The results in Table 1 do not account for the interaction of the $n = 0$ and $n = \pm 1$ modes. When the Reynolds number is increased to $\mathbb{R}_1 = 1.06$, these numerical results are robust. There is sensitivity to changes in a. Figure 4 shows the bifurcation diagram as the parameter ϵ_1 varies.

When a shifts away from 1.37, the modes shift away from being resonant (Figure 5(a) shows $a = 1.45$). Table 2 lists the conditions in which mode 1 is the critical mode and other modes are stable. As the annular gap widens, it is found that the critical mode shifts from 1 to 0. Figure 5(b) shows this for $a = 1.9$; mode 0 is already unstable at $\mathbb{R}_1 = 0.3$ for $0 < \alpha < 0.9$ while other modes are stable. Thus, Table 3 lists the conditions in which mode 1 is at onset, while mode 0 is already unstable due to capillary effects. When

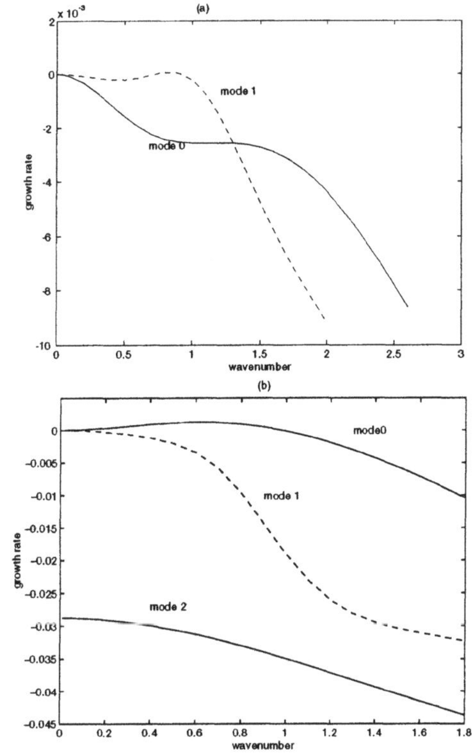

Figure 5. Growth rate versus wavenumber for $J = .01, m = .03, \zeta = 1.1, K = .10429$. (a) $a = 1.45$, $\mathbb{R}_1 = 0.6$; (b) $a = 1.9$, $\mathbb{R}_1 = 0.36$.

the Reynolds number $\mathbb{R}_1$ is increased, the added shear has a stabilizing effect on the capillary instability for mode 0 while mode 1 is destabilized. At $a = 1.9$, either mode 0 is unstable for lower Reynolds numbers or both modes 0 and 1 are unstable for higher Reynolds numbers, with mode 1 becoming more unstable than mode 0 (at $a = 1.9$, $\mathbb{R}_1 = 0.4$, mode 0 is unstable for $0.4 < \alpha < 0.9$ with maximum growth rate Re $\sigma = 0.0005$ at $\alpha = 0.7$, mode 1 is unstable for $0 < \alpha < 0.7$ with Re $\sigma = .002$ at $\alpha = 0.6$, thus mode 1 has the higher growth rate). In these cases where mode 1 dominates the instability at slightly higher Reynolds numbers, the nonlinear analysis at the neutral situations listed in Table 3 may be relevant. In both tables, the last columns denote the solution type: s denotes stable snakes, c denotes stable corkscrews. Both solution types are supercritical for $\epsilon_1 > 0$, the regime where the base flow is unstable, so that one type is stable. The trend is that snakes are preferred when the annular gap is wider.

a	$\mathbb{R}_1$	α	Im σ	β_1	β_2	pattern
1.37	1.0	1.3	-1.26	-6+53i	-25+680i	c
1.4	0.8	1.0	-0.98	-27+55i	-4-65i	s
1.45	0.6	0.8	-0.79	-41+17i	-1-19i	s
1.5	0.5	0.7	-0.70	-21+0.3i	-0.7-9i	s
1.6	0.43	0.6	-0.61	-5-0.2i	-0.5-3i	s
1.7	0.38	0.5	-0.51	-2+0.7i	-0.3-0.9i	s

TABLE 2. Critical conditions for $m = 0.03$, $\zeta = 1.1$, $K = 0.10429387$, $J = 0.01$. The last column shows whether the snakes or corkscrews are stable.

a	$\mathbb{R}_1$	α	Im σ	β_1	β_2	pattern
1.8	0.37	0.5	-0.51	-1+0.9i	-0.4-0.09i	s
1.9	0.36	0.5	-0.52	-0.6+0.9i	-0.5+0.2i	s

TABLE 3. Conditions for mode 1 to be neutral. Mode 0 onsets at a lower Reynolds number. $m = 0.03$, $\zeta = 1.1$, $K = 0.10429387$, $J = 0.01$. The last column shows whether the snakes or corkscrews are stable.

The base velocity profiles for the situations in Tables 2 - 3 are of the type shown in Figure 2(a). Note that since the dimensional centerline velocity $W_0^*(0) = (\rho_1 - \rho_2)gR_1^2A/[(K-1)4\mu_2]$, the sign of gravity is the same as the sign of $(1-\zeta)A/(K-1)$. Thus, for our parameters, g is positive and points in the same direction as the centerline velocity; this signifies down-flow. For $K = 0.10429$ and $\zeta = 1.1$, we see from $f^*/(\rho_1 g) = (1-\zeta K)/(K-1)$ that the applied pressure gradient f^* acts in the opposing sense to gravity and almost balances out the effect of gravity for the core fluid.

5.2. FREE FALL

In free fall ($f^* = 0$), $K = 1/\zeta$. We begin by considering the same fluid parameters as above ($m = 0.03, \zeta = 1.1, \mathbb{R}_1 = 1, \mathrm{J} = 0.01$), and find that mode 0 onsets first. The pattern selection analysis is performed at the wavenumber of maximum growth rate close to neutral situations for mode 1. In Table 4(i), $m = 0.03, \zeta = 1.1$, mode 0 attains the largest growth rate for $a = 1.4$ and mode 1 for $a = 1.5$ (the growth rate profile is similar to Figure 6(a)) and higher. At yet larger values of a, for example $a = 5$, only mode 0 is unstable. This is due to capillary effects for the relatively

	a	α	Im σ	β_1	β_2	pattern
(i) $\mathbb{R}_1 = 1$						
	1.4	2.8	-2.6	-2.2+40 i	-2.6 +38i	c
	1.5	2.2	-2.1	-1.9+27i	-2.1 + 25i	c
	1.6	1.8	-1.7	-1.67 + 18i	-1.71 + 15i	c
	1.7	1.6	-1.5	-1.44 + 13i	-1.35 + 10i	s
	1.8	1.4	-1.3	-1.16 + 9i	-1.04 + 7i	s
	1.9	1.3	-1.3	-0.97+7i	-0.78 + 5i	s
	2.0	1.2	-1.2	-0.8 +5i	-0.6+3i	s
	3.0	0.8	-0.8	-0.1 + 0.9i	0.03+0.5i	s
(ii) $\mathbb{R}_1 = 0.8$	a	α	Im σ	β_1	β_2	pattern
	1.2	4.6	-4.0	-0.8+32i	-1+30i	c
	1.3	3.4	-3.1	-1.4 +40i	-1.8 +39i	c
	1.4	2.6	-2.4	-1.6 +34i	-1.8 + 31i	c
	1.5	2.2	-2.1	-1.540 + 27i	-1.538 + 24i	s
	1.6	1.8	-1.7	-1.36 + 18i	-1.25 + 15i	s
	1.7	1.6	-1.5	-1.21 + 13i	-0.97 + 11i	s
	1.8	1.4	-1.3	-1.0 +9i	-0.8 + 7i	s
	1.9	1.3	-1.3	-0.9 + 7i	-0.6 + 5i	s

TABLE 4. Pattern selection analysis is performed at the maximum growth rate wavenumber for mode 1 at $m = 0.03, \zeta = 1.1, K = 1/\zeta, J = 0.01$, free fall.

thin filament at the core, since the shear rate at the interface scales small. Corkscrews are preferred unless the annulus width becomes large. A decrease in the Reynolds number eventually leads to mode 0 attaining the largest growth rate due to capillary instability. For the Reynolds number in Table 4(ii), for instance, mode 1 has the largest growth rates for $a = 1.6$ and higher, while mode 0 has the largest growth rates for the lower values of a. The growth rates here are however small, and we have applied the pattern selection analysis at the points of maximum growth rates. The results remain robust at the neutral situations; for example, the $a = 1.2$ entry in Table 4(ii) becomes neutral with $\alpha = 4.2$ at $\mathbb{R}_1 = 0.7$, where corkscrews are again stable. Calculations were performed for lower Reynolds numbers where mode 1 attains neutral conditions and, for example, at Reynolds number 0.6, the preferred patterns for $a = 1.3$ to 1.6 of Table 4(ii) remain the same. The results in Table 4 indicate that when there is more of the annular fluid, snakes are preferred while corkscrews are preferred when the

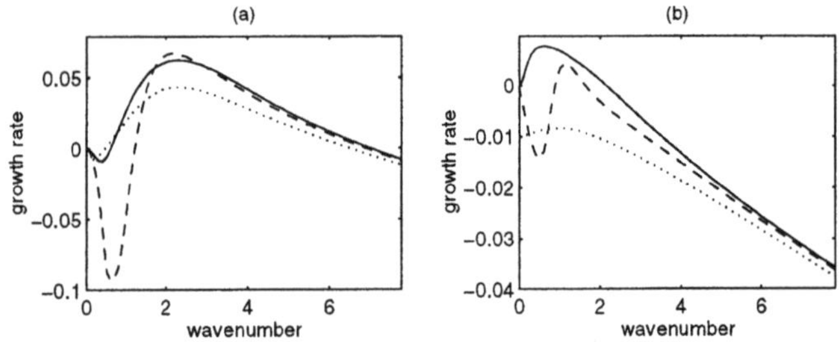

Figure 6. Growth rate versus wavenumber for free fall, $\zeta = 1.1$, $K = 1/\zeta$, $m = .001$, $\mathbb{R}_1 = 1$, (a) $a = 1.2$, (b) $a = 1.9$. Modes 0 (line), 1 (dash) 2 (dot).

annulus is narrow. This trend is also observed at other values of m, namely $m = 0.1, 0.001$. At $m = 0.1, \mathbb{R}_1 = 1$, mode 0 attains the largest growth rate for $a = 1.3$ to 1.5 while for higher $a = 1.6$ to 1.9, mode 1 attains the largest growth rate. The pattern selection analysis is performed at the largest growth rate wavenumber for mode 1. Again, the corkscrews are stable when the annulus is not too wide $a = 1.3$ to 1.7, and snakes are stable when $a = 1.8, 1.9$. At $m = 0.001, \mathbb{R}_1 = 1, 1.5$, mode 1 has the larger growth rate for the smaller $a = 1.2$ to 1.6 and mode 0 for the larger $a = 1.7$ to 1.9 (Figure 6). It is interesting that in the previous cases, the trend was the reverse. The pattern selection analysis for $\mathbb{R}_1 = 1$ at the largest growth rate wavenumber for mode 1 yields stable corkscrews for $a = 1.2$ to 1.5 and stable snakes for $a = 1.6$ and higher.

Next, consider free fall for equal viscosity and a heavier fluid outside: $m = 1, \zeta = 2, K = 0.5, \mathbb{R}_1 = 1$. The base flow for $a = 1.7$ is shown in Figure 2(b). This case is focussed on the effect of density stratification. For $a = 1.2$ to 3 in Table 5, only mode 1 is unstable. Modes 0 and 2 are stable (see Figures 7(a) - 7(b)), and in particular, mode 1 changes smoothly from Figure 7(a) to 7(b). At $a = 1.1$, all modes are stable. For the cases $a = 1.2$ to 3, mode 1 is unstable for long waves, and achieves a maximum growth rate at an order 1 wavenumber. The pattern selection results in Table 5 favor corkscrews unless the annulus is large. For larger annulus values, such as $a = 5$, mode 1 is stabilized and mode 0 is destabilized by capillarity. Calculations were also performed at the neutral stability points, e.g., $a = 1.9, \alpha = 1.2, \sigma = -0.0001 - i$, and again the preference for corkscrews is robust.

5.3. EXPERIMENTS OF BAI, CHEN AND JOSEPH (1992)

Figure 3 of Joseph *et al.*(1997) shows a variety of flow types in down-flow in a vertical pipeline experiment with oil and water, which was earlier re-

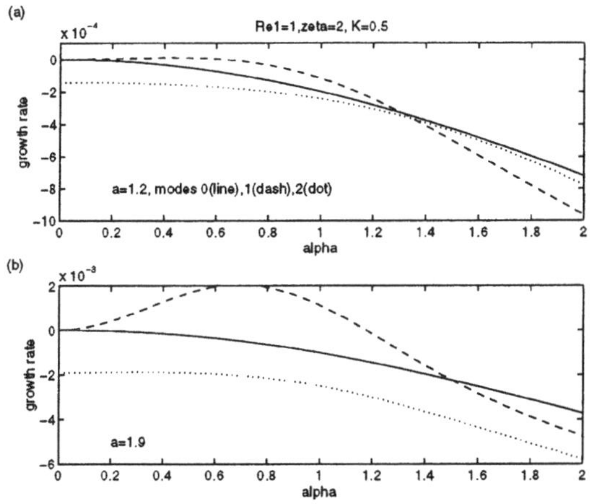

Figure 7. Growth rate versus wavenumber for free fall, $\zeta = 2, K = 0.5, m = 1, \mathbb{R}_1 = 1$. (a) a=1.2 (b) a=1.9.

a	α	Im σ	β_1	β_2	pattern
1.2	0.4	-0.15	-1.3-50i	-1.5-90i	c
1.3	0.7	-0.36	-0.8-24i	-1.1-30i	c
1.4	0.8	-0.49	-0.6-13i	-0.9 -9i	c
1.5	0.8	-0.55	-0.5-8i	-0.8-3i	c
1.6	0.8	-0.60	-0.4-5i	-0.7 -0.9i	c
1.7	0.8	-0.63	-0.4-4i	-0.7-0.1i	c
1.8	0.7	-0.58	-0.3-2i	-0.5-0.1i	c
1.9	0.7	-0.60	-0.3-2i	-0.4 -4i	c
2.0	0.7	-0.61	-0.25-1.5i	-0.36 -0.08i	c
2.1	0.6	-0.54	-0.18-0.9i	-0.24 -0.09i	c
2.3	0.6	-0.55	-0.16-0.6i	-0.19 - 0.1i	c
3.0	0.4	-0.39	-0.056-0.09i	-0.04-0.07i	s

TABLE 5. Pattern selection analysis is performed at the maximum growth rate wavenumber for mode 1 at $m = 1, \zeta = 2, K = 1/\zeta, J = 0.01, \mathbb{R}_1 = 1$, free fall.

ported in Bai *et al.*(1992). Flow type (c) in the regime called 'disturbed core annular flow' consists of corkscrew waves. There is some variation in the observed data, the conditions may not be exactly steady and the arrange-

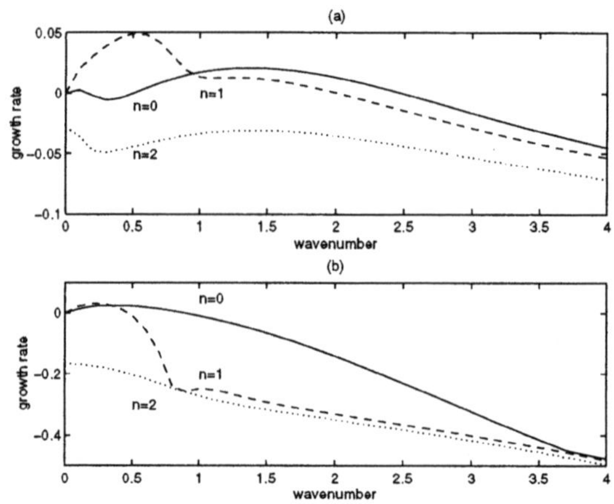

Figure 8. Growth rates Re σ versus wavenumber α for $K = -0.5427$, $a = 1.7$, $\zeta = 1.1$, $m = 0.00166$, $J = 0.06$. (a) $\mathbb{R}_1 = 1.2$ (b) $\mathbb{R}_1 = 0.2$.

ment may have narrower annuli. Snake waves were not observed. Further details are given in § 4 of Hu & Patankar (1995) where the parameters are $\mathbb{R}_1 = 1.2, a = 1.7, m = .00166, \zeta = 1.1, K = -.54270708$; their Fig. 2 shows that mode 1 has the largest growth rate at $\alpha = 0.531$ and mode 0 has much smaller growth rates, though it has two bands of unstable wavelengths, one for long waves due to interfacial tension and the other at O(1) wavenumbers due to shear. They show that the wavelengths and wavespeeds agree with the experimental values. Under these conditions, we need to examine what would happen if only the non-axisymmetric mode were excited. This could occur for instance, if the initial condition favors it. We can also cover a range of Reynolds numbers while keeping the relative driving forces K constant, to examine neighboring regimes. When the Reynolds number $\mathbb{R}_1$ is decreased, the O(1) wavenumber instabilities are stabilized since these are primarily due to shear, while the long-wave axisymmetric modes are destabilized due to capillary breakup. Thus, as the Reynolds number is decreased, the growth rates for mode 1 decrease, while that of the longer waves at mode 0 increase and overtake them at approximately $\mathbb{R}_1 = 0.2$. This is illustrated in Figure 8. We show the pattern selection results in Figure 9 for $0.2 < \mathbb{R}_1 < 1.4$ which includes the parameter values of interest. The base velocity profiles is of the type in Figure 2(c). Figure 10 shows the pattern selection results at another wavenumber $\alpha = 0.4$ where mode 1 attains largest growth rates for the lower Reynolds numbers, showing the preference for corkscrews around $\mathbb{R}_1 = 1.2$. It is interesting that our bifurcation analysis shows this and that snake waves were not observed in the

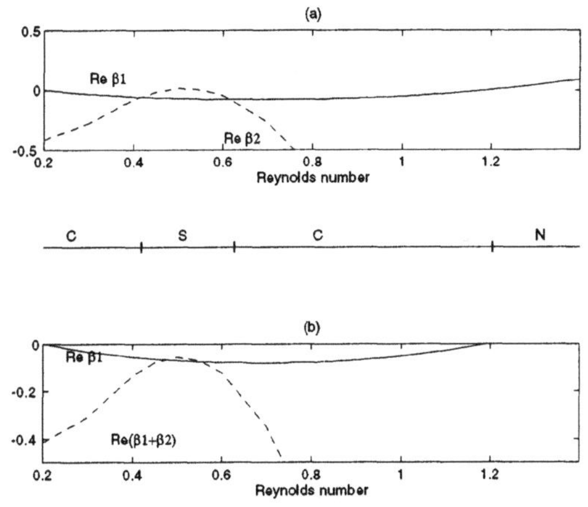

Figure 9. Pattern selection results for $K = -0.5427$, $a = 1.7$, $\alpha = 0.531$, $\zeta = 1.1$, $m = 0.00166$, $J = 0.06$. (a) This shows the real parts of the Landau coefficients (cf. Eqs.(11), (13)); a stability condition for corkscrews is that Re $\beta_2 <$ Re β_1, and vice versa for snakes. (b) The second stability condition, that of supercriticality, for corkscrews is that Re $\beta_1 < 0$, and for snakes that Re $(\beta_1 + \beta_2) < 0$. The results from both (a) and (b) are combined in the middle plot to show the intervals of Reynolds numbers $\mathbb{R}_1$ in which corkscrews (C) are preferred, snakes (S) are preferred, or neither (N).

experiments.

Acknowledgements

This work was supported by the National Science Foundation under Grant No. CTS-9307238 and CTS-9612308 and the Office of Naval Research under Grant No. N00014-92-J-1664.

References

BAI, R., CHEN, K., & JOSEPH, D. D. 1992 Lubricated pipelining: stability of core-annular flow. Part 5. Experiments and comparison with theory, *J. Fluid Mech.* **240**, 97-132.

BOOMKAMP, P. A. M. & MIESEN, R. H. M. 1992 Nonaxisymmetric waves in core-annular flow with a small viscosity ratio, *Phys. Fluids A* **4**, 1627-1636.

FUJIMURA, K., & RENARDY, Y. Y. 1995 The 2:1 steady/Hopf mode interaction in the two-layer Benard problem, *Physica D* **85**, 25-65.

GOTTLIEB, D., & ORSZAG,S. A. 1977 *Numerical analysis of spectral methods: theory and applications*, CBMS-NSF Regional conference series in applied mathematics. Society for Industrial and Applied Mathematics, Philadelphia.

HU, H.H., & PATANKAR, N. 1995 Non-axisymmetric instability of core-annular flow, *J. Fluid Mech.* **290**, 213-224.

JOSEPH, D. D., RENARDY, Y., & RENARDY, M. 1983 Instability of the flow of immiscible liquids with different viscosities in a pipe, *J. Fluid Mech.* **141**, 309-317. Longer version available as University of Wisconsin-Madison Mathematics Research Center Report.

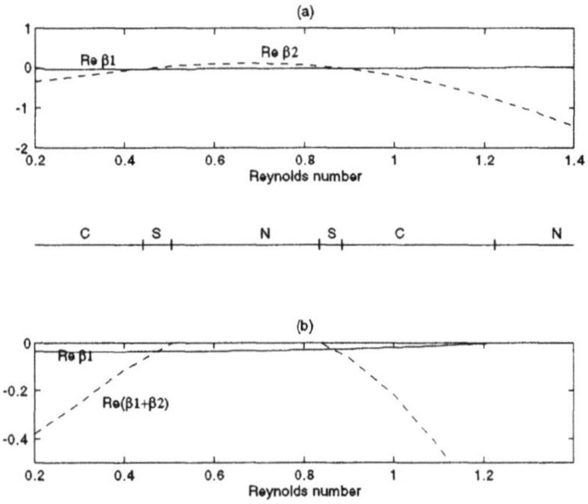

Figure 10. Pattern selection results for $K = -0.5427$, $a = 1.7$, $\alpha = 0.4$, $\zeta = 1.1$, $m = 0.00166$, $J = 0.06$. (a) This shows the real parts of the Landau coefficients (cf. Eqs.(11), (13)); a stability condition for corkscrews is that Re $\beta_2 <$ Re β_1, and vice versa for snakes. (b) The second stability condition, that of supercriticality, for corkscrews is that Re $\beta_1 < 0$, and for snakes that Re $(\beta_1 + \beta_2) < 0$. The results from both (a) and (b) are combined in the middle plot to show the intervals of Reynolds numbers $\mathbb{R}_1$ in which corkscrews (C) are preferred, snakes (S) are preferred, or neither (N).

JOSEPH, D. D., & RENARDY, Y. Y.(1993)*Fundamentals of Two-Fluid Dynamics, Part 1: Mathematical Theory and Applications, Part 2: Lubricated Transport, Drops and Miscible Liquids* Springer Verlag New York.

JOSEPH, D. D., BAI, R., CHEN, K., & RENARDY, Y.(1997)Core-Annular Flows, *Ann. Rev. Fluid Mech.* vol. 29. pp.65-90.

ORSZAG, S. A.(1971)Accurate solution of the Orr-Sommerfeld stability equation, *J. Fluid Mech.* **50**, 689-703.

PREZIOSI, L., CHEN, K., & JOSEPH, D. D.(1989)Lubricated pipelining: stability of core-annular flow, *J. Fluid Mech.* **201**, 323-356.

RENARDY, Y. (1989) Weakly nonlinear behavior of periodic disturbances in two-layer Couette-Poiseuille flow, *Phys. Fluids A* **1**, 1666-1676.

RENARDY, Y. (1997)Snakes and corkscrews in core-annular down-flow of two fluids, *J. Fluid Mech* in press.

RENARDY, M., AND RENARDY, Y. (1993)Derivation of amplitude equations and analysis of sideband instabilities in two-layer flows, *Phys. Fluids A* **5**, 2738 - 2762.

RUELLE, D. 1973 Bifurcations in the presence of a symmetry group, *Arch. Rat. Mech. Anal.* **51**, 136.

ANALYSIS OF A FREIGHT PIPELINE SYSTEM

T.S. Lundgren & Y. Zhao
Dept. Aerospace Engineering and Mechanics
University of Minnesota
Minneapolis, MN 55455

ABSTRACT: This paper examines the aerodynamic characteristics of a freight pipeline system in which freight capsules are individually propelled by electrical motors. The fundamental difference between this system and the more extensively studied pneumatic pipeline is the different role played by aerodynamic forces. In a driven system the propelled capsules are resisted by aerodynamic forces and, in reaction, pump air through the tube. In contrast, in a pneumatic system external blowers pump air through the tubes and this provides thrust for the capsules. An incompressible transient analysis is developed to study the aerodynamics of multiple capsules in a cross linked two-bore pipeline. An aerodynamic friction coefficient is used as a cost parameter to compare the effects of capsule blockage and headway and to assess the merits of adits and vents. We conclude that optimum efficiency for off-design operation is obtained with long platoons of capsules in vented or adit connected tubes.

1. Introduction

Pipeline systems for freight transportation have many desirable features. Freight pipelines buried underground would have little environmental impact on surroundings once installed. These systems can be fully automated and do not interfere with human movement. Pipeline systems are closed and can thus be operated regardless of weather conditions.

Freight pipeline systems that have been extensively studied use pneumatic blowers as the power source. They move capsules through a duct by using a vacuum or air pressure. Booster pumps have to be used to transport cargos beyond a distance of several miles. Since capsules cannot go through the pumps, it becomes necessary to interpose a set of valves and airlocks through which the capsules can bypass the booster pumps. In addition, pneumatic pipelines suffer from low energy efficiency and high noise level. Coupled with the fact that tunneling technologies were very expensive and challenging until recently, these system concepts were considered impossible for general freight and were dubbed "pipe dreams." Although the economics of pneumatic pipelines appear favorable (Round and Marcu, 1986) they have not been widely used except for transporting coal and other bulk materials over short ranges.

Several electrical propulsion concepts for pipeline systems have been proposed. Ampower Corporation (1984) proposed a capsule pipeline system powered by linear induction motors. Economic analyses were conducted based on experience to demonstrate the feasibility of such a concept. However, the technical details are yet to be developed. Recently, Japanese engineers have constructed and tested a prototype tube transportation system powered by linear synchronous motors (Fujisawa & Tomoji, 1994).

1.1. DESCRIPTION OF FREIGHT TRANSPORT SYSTEM

In this paper we study the aerodynamic forces on a capsule in an electrically driven freight pipeline system. Specifically, we consider a system consisting of two parallel tubes each carrying one-way traffic of loaded capsules. There are some economic advantages to connecting these tubes at intervals by short cross tunnels, or adits (Gawthorpe, 1983), to allow pressure relief. The channel tunnel (Henson and Bradbury, 1991) is an example of an existing long two-bore system with adits. A schematic layout of such a system is shown in figure 1a.

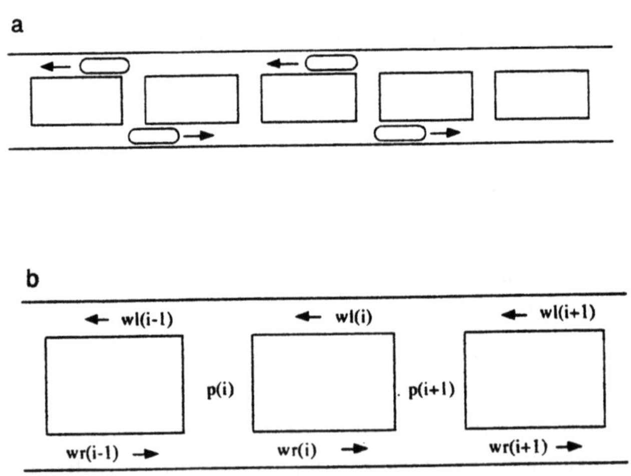

Figure 1. Layout of pipeline system showing left and right tubes and adits.

The idea behind adits is to allow the high pressure difference that builds up across a moving capsule to divert air through a cross tunnel into the other tube, thus producing circulation around an adit loop (consisting of both tubes and two successive adits) and limiting the length of air column a capsule must push. This produces air flow in the other tube in the direction of traffic flow and assists the motion of capsules there. Under ideal operating conditions in which there is uniform capsule traffic in each direction, each capsule with the same velocity and load and with the same headway between each, the cross tunnels have no effect. Each capsule pushes its own measure of tube air and there is no pressure imbalance to induce flow through the cross tunnels. However, when there is light or intermittent traffic the added flexibility in allowing air to be routed through a path of less resistance can reduce the drag. As an extreme example, if there were only one tightly fitting capsule in a long tube, with no adits or air vents for relief, the capsule would have to push the entire column of tube air against the frictional resistance of the tube walls. One purpose of this paper is to develop a method for assessing the effectiveness of adits and vents.

There are some negative aspects to the use of adits. When the capsule blockage is fairly large, very high tube air speeds can result, nearly as large as the capsule velocity, and there is a potential for high speed flow through the cross tunnels to produce damaging jet flow into the other tube. The side force could derail an unrestrained capsule or damage loose freight. Bernard (1993) reports that in the channel tunnel this possibility necessitated design changes in the cross tunnels and limitations on shuttle speeds.

Blockage is defined as the ratio of the cross sectional area of the capsule to the area of the tube. Blockage is a critical design parameter. Since tunneling is expensive the first thought is to make the capsule of circular section and as large as possible for a given tube diameter. However general freight is often palletized into rectangular units, which would leave packing voids in circular capsules. This suggests the use of capsules of square section. These would have smaller blockage, since the voids would then be on the outside. As an added advantage square capsules could be more conveniently stacked onto trucks for final destinations and the opportunity of extending vanes into the voids could give added drag control for braking (Tsugi et al, 1985). A square section capsule which just fits a circular tube has a blockage of .64. If some additional clearance were allowed at corners a blockage of .5 is realistic. (This is the blockage of a channel shuttle.) On the other hand some systems require large blockage. Some linear synchonous motors which have the electrical windings around the outside of the tube must necessarily have a small gap between the tube wall and permanent magnets carried on the capsule. The concept proposed by Ampower has a square capsule with circular seals at the ends which nearly block the tube, the purpose being to prevent capsule encounters by the compression of headway air. This would preclude the use of adits, since leakage through the adits would defeat the compression buffering.

The tube diameter itself is also an important parameter, which must be kept as small as possible to minimize boring costs. For general freight transport it would be difficult to have a diameter smaller than about two meters; we adopt this value when it is necessary to be specific about size. One might consider tubes of smaller diameter for parcel or document delivery. The methods developed here are applicable to any size tube.

2. Analysis of a Tube Freight Transport System

Consider a two-bore tube freight system consisting of parallel tubes, connected at regular intervals by pressure relief tunnels. The capsules are driven automatically by electric motors. The moving capsules cause flow in the tubes and through the adits. In the analysis we assume incompressible turbulent flow, and assume that the capsule length is short compared to the distance beween adits. The drag on a capsule is given by

$$D = .5\rho(V-w)^2 \mathrm{sgn}(V-w) A C_D \tag{1}$$

Where V is the capsule speed, w the velocity of the air in the tube and ρ is the gas density. The sgn function indicates the algebraic sign; if w is greater than V the drag becomes thrust. An expression for the drag coefficient C_D is derived in Lundgren & Zhao (1997). The simple result, similar to that of Tsugi (1985), is

$$C_D = \frac{\beta}{(1-\beta)^2} + \frac{(1+\beta^{1/2})}{(1-\beta)^3}\frac{4l}{d}C_{fg} \tag{2}$$

where $\beta = A/A_T$ is the blockage ratio, A_T the tube area, l and d the length and diameter of the capsule, and C_{fg} is a friction coefficient which is a function of the Reynolds number of the turbulent flow in the gap between the capsule and the tube. The friction coefficient is given approximately by the Blasius formula (Schlichting, 1979):

$$C_{fg} = .316/4R_{eg}^{1/4}, \quad R_{eg} = \frac{2(R_T - R)v}{\nu} = \frac{2R_T(V-w)}{\nu(1+\beta^{1/2})}. \tag{3}$$

The first term in eq.(2) is a large pressure drag caused by separation at the rear of an (assumed) unstreamlined capsule. The second term is a viscous effect caused by the flow in the gap. The drag coefficient depends on the relative velocity $V-w$ through the friction coefficient. When β is near one (full blocking) the drag coefficient can be 1000 or more. This does not necessarily mean that the drag itself is large because $V - w$ can become small enough to compensate for the large C_D.

The pressure drop across a capsule is seen by the air in the tube as a pressure rise in the direction of motion and pushes the air through the tube as if it were produced by a pump. When there are a number of capsules in a bore section, each with different velocities in general, the sum of the pressure rises causes flow in each bore according to the following momentum equation; the rate of change of air momentum must equal the sum of the forces acting on the air. That is,

$$\frac{d}{dt}(\rho A_T\ L_T\ wr_i) = \sum_{ith\ sect.,r} .5\rho(V_j - wr_i)^2 \mathrm{sgn}(V_j - wr_i) A C_D(V_j, wr_i)$$

$$-.5\rho\ wr_i^2 \mathrm{sgn}(wr_i)\ 2\pi R_T\ L_T\ C_f(wr_i) + A_T(p_i - p_{i+1}). \tag{4}$$

$$\frac{d}{dt}(\rho A_T\ L_T\ wl_i) = \sum_{ith\ sect.,l} .5\rho(V_j - wl_i)^2 \mathrm{sgn}(V_j - wl_i) A C_D(V_j, wl_i)$$

$$-.5\rho\ wl_i^2 \mathrm{sgn}(wl_i)\ 2\pi R_T\ L_T\ C_f(wl_i) - A_T(p_i - p_{i+1}). \tag{5}$$

Here subscript i indicates the ith bore section, wr_i and wl_i are the air velocities in the right and left bores, subscipt j refers to the jth capsule and $C_D(V, w)$ is given by eqs.(2) and (3) with the dependence on the velocities indicated by the notation. The sums are over all the capsules in the appropriate bore of the ith section, which has length L_T. The friction coefficient $C_f(w_i)$ is again given by the Blasius formula but with Reynolds number based on the tube air velocity;

$$C_f(w_i) = .316/4R_{ei}^{1/4}, \quad R_{ei} = 2R_T w_i/\nu. \tag{6}$$

The variables p_i and p_{i+1} are the pressures at the ends of the ith bore section, assumed constant through a lossless adit. Figure 1b shows some of the notation.

The tube velocities must conserve mass at the junctions of bore sections with adits. Flow through an adit which decreases the flow rate in the right bore must be accounted for as an increased flow rate in the left bore. This may be expressed as

$$wr_i - wl_i = wr_{i+1} - wl_{i+1}. \tag{7}$$

The negative signs occur here because the wl variables are defined to be positive for flow to the left (in the direction of traffic there). There is a condition like eq.(7) for each adit.

The equations of motion of the individual capsules are

$$\frac{dx_j}{dt} = V_j, \tag{8}$$

$$M_j \frac{dV_j}{dt} = F_j - .5\rho(V_j - w_i)^2 \mathrm{sgn}(V_j - w_i) A C_D(V_j, w_i) - s(x_j) g M_j \tag{9}$$

where w_i refers to either wr_i or wl_i depending on which bore the jth capsule is in. Here F_j is the force applied to the jth capsule by the motor, $s(x_j)$ is the local slope of the tube bed and g the gravitational constant. In these equations x_j and V_j are taken as positive in the direction of traffic in the appropriate bore. This set of equations is strongly coupled since the drag force on the jth capsule depends on w_i for the section and bore it is in, and w_i in turn depends on the velocities of all the capsules in the system, the coupling being through the pressure variables.

In Lundgren & Zhao (1997) this set of equations is put in a form which may be solved numerically and a numerical method is described. This involves using the conditions expressed by eq.(7) to solve for the pressure variables as functions of capsule and tube velocities. The numerical problem is considerably simplified if instead of adits we have simple lossless vents to atmosphere at the adit locations. Then we can simply omit the pressure variable terms from eqs.(4) and (5).

3. Design Conditions

Under optimum operating conditions there is balanced traffic in each bore. An equal number of capsules going in each direction, each with the same load and velocity and with the same spacing between them. Under such "design conditions" there are the same number of capsules in each bore section and therefore w_i is the same for each. The basic parameters which need to be specified to design the system are;

V_d, the design velocity,

q_d, the number of capsules per sec. ($1/q_d$ is the headway time, V_d/q_d the headway length),

M_d, the design mass of capsule plus payload,

A_T, the tube area.

In addition details of the capsule need to be specified, such as the capsule area A and the length to diameter ratio, l/d.

The steady versions of eqs.(4) and (9) give F_d, the motor force required to maintain the speed of a capsule, and w_d, the common tube velocity. It is not necessary to specify L_T, the distance between adits, at this stage because the number of capsules per unit length is q_d/V_d, the reciprocal of the headway length, and therefore the number of capsules in a bore section is $L_T q_d/V_d$. Because of these considerations L_T factors out of eq.(4), resulting in

$$.5\rho(V_d - w_d)^2 A C_{Dd}(q_d/V_d) = .5\rho w_d^2 2\pi R_T C_{fd}. \tag{10}$$

Here we have used $C_{Dd} = C_D(V_d, w_d)$ and $C_{fd} = C_f(w_d)$. Solving this equation for w_d gives

$$\frac{w_d}{V_d} = \frac{\phi}{1+\phi} \tag{11}$$

where ϕ is defined by

$$\phi = \left(\frac{\beta R_T q_d C_{Dd}}{2V_d C_{fd}}\right)^{1/2}. \tag{12}$$

In these expressions C_{fd} and C_{Dd} depend on w_d through the Reynolds numbers, so it is necessary to iterate numerically in order to solve for w_d and ϕ. The parameter ϕ represents the relative effects of drag and wall friction. When C_{Dd} is very large, ϕ is large and w_d can become nearly as large as V_d, while if β is small, ϕ and therefore w_d are small.

Equation(9) gives

$$F_d = .5\rho(V_d - w_d)^2 A C_{Dd}$$

or with w_d from eq.(11),

$$F_d = .5\rho V_d^2 A_T \beta C_{Dd}/(1+\phi)^2. \tag{13}$$

The quantity F_d is required in order to design and size the electrical propulsion system.

There are two results which can be derived here from these simple equations.

3.1. COMPARISON WITH A PNEUMATIC SYSTEM

In a pneumatic system air is forced through the tubes by external pumps to drive the capsules, while for the motor driven system air is forced by the motion of the capsules themselves. In the most favorable pneumatic case blockage is large and the air velocity is equal to the capsule velocity V_d; there is no drag or thrust on the capsule. (In a less favorable scenario the air velocity would have to be greater than V_d, passing over the capsule from behind to provide enough thrust to overcome rolling friction.) The pneumatic pumps would have to produce a pressure force

$$.5\rho V_d^2 2\pi R_T C_f(V_d) V_d/q_d \tag{14}$$

for each capsule in order to overcome friction in the headway length between capsules. This force should be compared with F_d. The ratio of the latter to the former is the efficiency, η, of the driven system relative to a pneumatic system with the same capsule velocity and tube diameter:

$$\eta = \left(\frac{\phi}{1+\phi}\right)^{7/4}. \tag{15}$$

The fractional exponent results from different velocities in the two friction coefficients. Clearly η is less than one, so the driven system always requires less power. As $\beta \to 1$ the two systems become equivalent, but at $\beta = .5$ there is nearly a 50% improvement.

3.2. OPERATING COST UNDER DESIGN CONDITIONS

A useful cost parameter can be defined as the work required to transport one unit of weight a distance of one unit of length. The work required to move a capsule a distance L (against aerodynamic forces only) is $F_d L$. This should be divided by the ***payload*** weight, gM_p, times the distance L to get an aerodynamic cost parameter, μ:

$$\mu = F_d/gM_p \ . \tag{16}$$

This dimensionless quantity is effectively a friction coefficient, but it should be kept in mind that the actual cost is independent of the payload; it costs the same to move an empty capsule. This is a useful parameter for comparison with other modes of freight movement, since it can be converted to a dollar cost per ton-mile. The cost parameter μ is plotted versus β in figure(2) for the situation described in the following paragraph.

We have adopted Vandersteel's model (Ampower Corporation, 1984) which has a 2m diameter tube carrying 10 ton capsules (of which 8 tons are payload), propelled at 25 m/s at 2 sec intervals. This gives a continuous freight rate of 4 tons/sec, or 100 million tons/yr, which is about 10 times the annual truck freight tonnage carried between Minneapolis and Chicago, but this is not an unreasonable rate for some heavily used corridors. In the figure we show results for single capsules and for short trains of coupled capsules. We have considered 3-capsule trains at 6 second intervals and 5-capsule trains at 10 second intervals. These both transport 4 tons/sec but have greater spatial headway between units. We also show the same 3 cases with 50 m/s capsule velocity. We observe, of course, that higher velocity costs much more. For both velocities there is some cost advantage for short trains over single capsules if the blockage is not too great. This is because the pressure drag is the same for a train as it is for a single capsule and this distributes the pressure drag tare over more capsules (this is effectively a form of streamlining), however as the blockage increases this becomes irrelevant since most of the resistance is then due to air friction in the headway space between trains.

For comparison purposes other sources of resistance are rolling friction, for which friction coefficients are of order .001, and gravity, where a one percent grade offers a resistance coefficient of .01. (In this paper we have considered only level operation; gravity is obviously an important consideration in general.)

The cost parameter can be put in monetary terms by converting first to ft-lbs/ton-mi and then converting ft-lbs to kw-hrs, finally assuming 7 ¢ per kw-hr (the electrical rate in Minneapolis) we get a conversion factor of 28: the cost in ¢/ton-mi = 28 μ. A cost parameter of .01 on the figure corresponds to .28 ¢ /ton-mi. This cost estimate neglects rolling friction and gravity and assumes perfect conversion of electrical energy; the figures seem low because capital costs are not considered here.

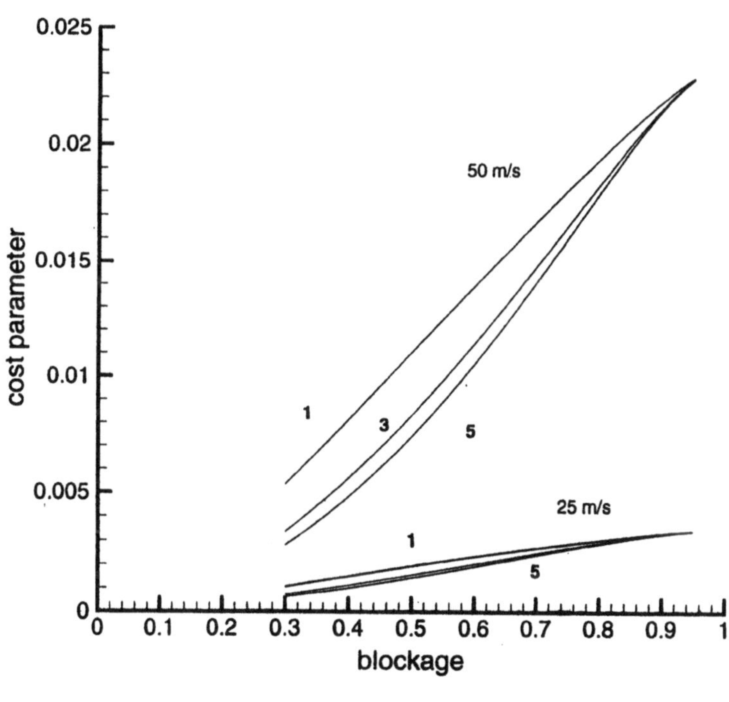

Figure 2. Cost parameter versus blockage for 1, 3, and 5 capsule trains at two velocities

4. Off-Design Operation: The Case for Adits and Vents

The simple economics of tube freight transport are quite favorable at the tonnage rates (4 ton/sec) used in the above estimates. However, it is inevitable that there will be periods of light traffic when it is not possible to inject fully loaded capsules into the pipetube at the design headway interval. There are two operating strategies one can consider: intermittent operation, in which loaded capsules are inserted as they become available, and platoon operation in which capsules are held back at the warehouse until there is a sizeable number, which are then each inserted at the design headway forming a compact group, to be followed by a, perhaps lengthy, period while a sufficient number are collected for the next platoon. We find below that these strategies are equivalent when there are no adits or vents; with adits the platoon strategy is much superior.

4.1. INTERMITTENT OPERATION

We consider the cost of transporting the standard 10 ton capsules, with 8 ton payloads, without adits, in an off-design situation in which the capsules are evenly spaced but the headway between the capsules is greater than the 2 second design value. We consider two blockage values, $\beta = .5$ and $\beta = .9$. It is assumed that the motor force is available to keep the speed of each capsule at 25 m/s. In figure 3 the cost parameter is plotted versus headway time for the range 2 seconds to 20 seconds, i.e., the tonnage rate varies from 4 ton/sec on the left side of the figure to .4 ton/sec on the right. The solid curves without symbols are without adits or vents. (Results for adits are shown with solid symbols; vents with open symbols.) For $\beta = .9$ the cost increases by a factor of 9 over this range, while for $\beta = .5$ it increases by a more moderate factor of 4. At the higher blockage value the cost increases almost linearly with headway time, most of the resistance coming from air friction in the linearly lengthening distance between capsules.

When adits (or vents) are considered it is necessary to specify both the adit spacing and the total length of the pipeline. We have computed for a 20 km pipeline with 80 adits; the adit spacing is 250 m. Under design conditions the spatial headway is 50 m, so there would be 400 capsules in each tube and 5 capsules between adits in each tube. As we increase the headway between capsules there is no effect on cost as long as there is at least one capsule in each adit section. That is, no effect up to a headway time of 10 sec, and this is also the case if there is no traffic in the other tube; at longer headways we begin to see an effect. In the figure the symbols are obtained by using the computational method described by Lundgren & Zhao (1997), with one tube empty. The results for adits, shown as solid symbols, are disappointing. For $\beta = .5$ there is no effect over the range shown, while for $\beta = .9$ there is only a small improvement. When the capsules are spaced out like this there is only a small opportunity for air to be shunted through the other tube.

For simple lossless vents to the atmosphere, at the same spacing as for the adits, the results (open symbols) are somewhat better for the larger blockage, but still have no effect for $\beta = .5$.

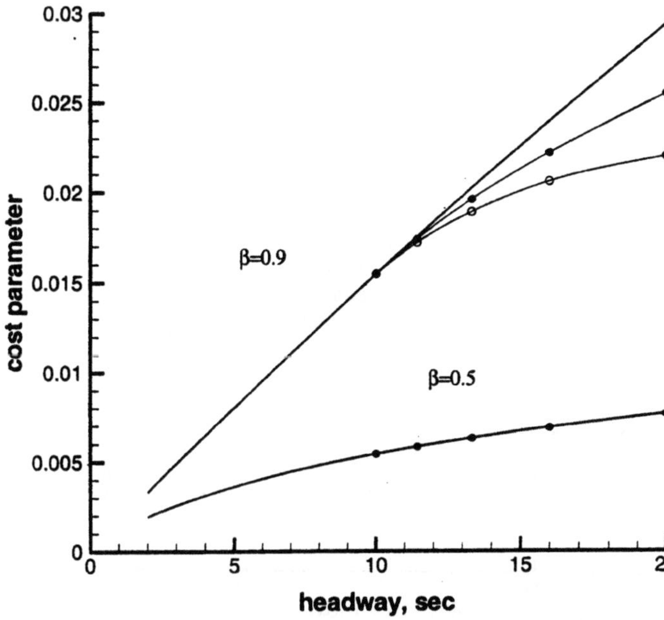

Figure 3. Cost parameter versus headway, when capsule spacing is greater than the design value. Solid circles are for adits, open circles for vents.

4.2. PLATOON OPERATION

We consider platoons of from one to 80 capsules, with 25 m/s speed and 2 sec headway in a 20 km pipeline with 250 m between adits or vents. The other tube is empty. The results are shown in figure 4 as relative cost versus platoon length for $\beta = .5$ and $\beta = .9$, solid symbols for adits, open symbols for vents. Relative cost is the cost divided by the cost under design conditions. If the platoon length were increased to 400 for this pipeline the relative cost would be one. If the capsules in a 40 capsule platoon were spaced out evenly they would have headway of 20 seconds; the situation would be the same as the right hand side of figure 3. The advantage of platoon operation over spaced operation are clear from these figures. The physical reason is that with a platoon it is not necessary to pump air through the entire tube; the adits work as intended, circulating the air through the other tube. Vents work even better.

It is interesting that the *relative* results are almost independent of blockage. Smaller blockage is cheaper, however, because this is the case for design conditions. From figure 3 the design cost parameter (2 sec. headway) for $\beta = .5$ is .0019, while it is .0033 for $\beta = .9$, a factor of about 1.7 larger.

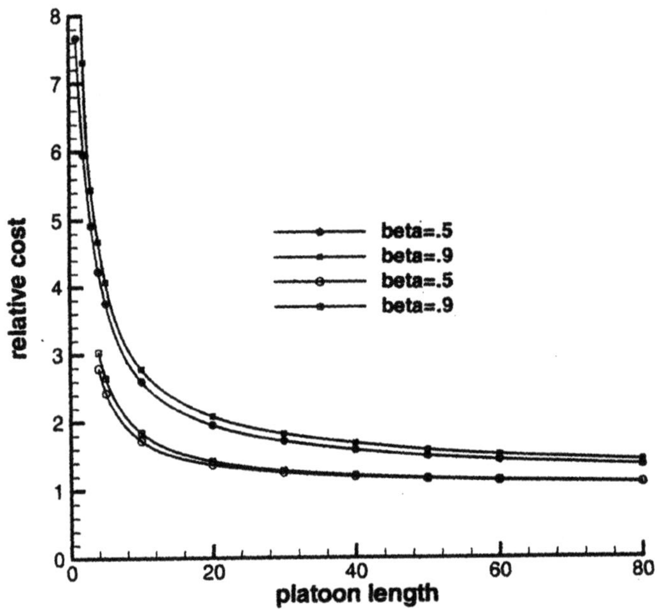

Figure 4. Cost relative to design values versus number of capsules in a platoon. Solid circles are for adits, open circles for vents.

5. Conclusions

In this paper we have described an analysis, and computational method, for use in the design of underground freight pipeline systems. (This analysis is described in more detail by Lundgren & Zhao, 1997). The methods can be used for two-bore systems, of any length and diameter, connected at intervals by adits or vented at intervals to atmosphere. Our examples have centered on 2 m diameter tubes, a size which is suitable for general freight, carrying 10 ton capsules at the relatively low speed of 25 m/s with 2 sec headway. When the tubes are operating at capacity we refer to this as "design conditions" and relate off-design operational costs to this ideal.

It is shown that there is an intrinsic advantage of this kind of system over pneumatic systems if the capsules have moderate blockage, since some of the air set in motion by the capsules can pass over the capsules. For a blockage of .5, which can be achieved without making the tube larger by using capsules of square section, the aerodynamic cost is about half that for pneumocapsules.

Potential cost problems arise when the system is not operating at full capacity; the aerodynamic cost per ton-mi would be 4 to 9 times the optimum if the traffic were at 10% of capacity. This situation is much improved if the tubes are connected by adits or are vented, provided the traffic management strategy is to run platoons of capsules at the design headway, followed by long intervals with no traffic. If the physical length of the platoons is much longer than the spacing between adits the cost per ton-mi is only 50% greater than the optimum, and only 10% greater if vents are used instead. The advantage comes from circulating the driven air through the other tube, or diverting it to atmosphere, eliminating the necessity of pushing it at high speed against wall friction all the way through the tubes.

Acknowledgements

This research is supported by the Minnesota Department of Transportation. We thank the following individuals for many useful suggestions and discussions: John Sampson, Larry Foote, Greg Paul Busacker at Mn/Dot Environmental Services, Richard Lambert at Mn/Dot Office of Railroads & Waterways, Don Botz at Mn/Dot Office of Research Administration, Prof. D. D. Joseph at the University of Minnesota, and Mr. W. Vandersteel at Ampower, Inc.

References

Ampower Corporation (1984), "Pneumatic Capsule Pipeline System," U.S. Patent No. 4,458,602.

Bernard, K. (1993) " Eurotunnel - design and implementation" High Speed Ground Transportation Systems I: Proc. First Int. Conf. on High Speed Ground Transportation (HSGT) Systems. ed. Bondada, M.V.A. and Wayson, R.L., ASCE N.Y., 405–416.

Fujisawa, T., Misago, T. Araki, O. & Goriki, T. (1994) "Development of Linear Tube Transportation System," *NKK Technical Review*, **70**, 25-32. (Available from National Technical Information System, Springfield, VA)

Gawthorpe, R. G., Pope, C.W. & Green, R.H. (1979), "Analysis of train drag in various configurations of long tunnels," Proc. 3rd Int. Symposium on the Aerodynamics and Ventilation of Vehicle Tunnels, BHRA Fluid Engineering, Cranfield,Bedford, England, 257-280.

Hammitt, A. G. (1972), "Aerodynamic Analysis of Tube Vehicle Systems," *AIAA Journal*, **10**, 1972, 282-290.

Henson, D.A. and Bradbury,W.M.S. (1991),"The aerodynamics of channel tunnel trains", Int. Symp. on Aerodynamics and Ventilation of Vehicle Tunnels (7th: 1991: Brighton, England) ed. Haerter, A. , 927–956. Elsevier.

Round, G.F. & Marcu, M.I. (1987),"Pneumocapsule pipelines: Potential for North America," *J. Pipelines*, **6**, 221-238.

Schlichting, H., (1979), *Boundary Layer Theory*, 7th edition, McGraw-Hill.

Tsuji, Y. (1985), "Fluid Mechanics of Pneumatic Capsule Transport," *Bulk Solids Handling*, **5**, 653 - 661.

Tsuji, Y., Morikawa, Y., and Seki, W. (1985), "Velocity Control in a Capsule Pipeline by Changing the Area of the End-Plate," *J. Pipelines*, **5**, 147-153.

Lundgren, T.S. & Zhao, Y. (1997), "Aerodynamic Characteristics of an Electrically Driven Freight pipeline System," Submitted to J. Transportation.

PERISTALTIC TRANSPORT IN A FINITE CIRCULAR PIPE

H. RAMKISSOON
Department of Mathematics & Computer Science
The University of the West Indies
St. Augustine, Trinidad (W.I.)

L.K. Antanovskii
Modflow International Dty Ltd
259-261 Colchester Road
Kilsyth, Victoria 3/37
Australia

A viscous flow in a long pipe of a circular cross-section, driven by a time-dependent pressure drop at the ends of the pipe and the deformation of the pipe walls, is investigated. This kind of a flow is relevant to the peristaltic transport of blood, with the effect of the oscillating pressure gradient due to the heart pumping included. An explicit solution is obtained within the framework of lubrication theory approximation, and some numerical simulations to illustrate the model are carried out. In order to account for inertia effects, an approximation based on an a priori parabolic flow pattern across the pipe, is also derived.

1. Introduction

The word peristalsis is derived from the Greek word "PERISTALTIKOS" which means clasping and compressing. It is a major mechanism to transport liquid or solid substances in the living body. Intestinal peristalsis, for example, consists of a wave of contraction travelling along the gut and propelling its contents. It consists in a narrowing and transverse shortening of a portion of the tube which then relaxes while a lower portion becomes shortened and narrowed.

Within recent times there has been some interest in the peristaltic flow of fluids resulting in a number of research publications [1-12]. In particular, asymptotic analyses of peristaltic transport with respect to a small parameter, such as the radius-to-wavelength ratio (slenderness), were implemented in [7,12] at low Reynolds number, using the approach of lubrication theory.

However, to the authors' best knowledge, there is no study of peristaltic flow through a deformable pipe, subject to time-dependent variation of pressure at the pipe ends. This situation is typical for the flow in a blood vessel due to the heart pumping, and has application in medical sciences. In fact, this work has been motivated by a claim [10, 13] that the arteries exhibit peristaltis.

In this paper the lubrication theory approach is applied to a viscous flow in a finite pipe, subject to time-dependent pressure drop and deformation of the pipe walls. Unlike [7], the shape evolution of the pipe can be arbitrary except that the characteristic wave-length is assumed to be essentially greater than the average radius of the pipe cross-section.

2 Governing equations

Consider a long pipe

$$D = \{0 \le r < a(x,t),\ 0 \le x \le L\}$$

with deformable walls

$$S = \{r = a(x,t),\ 0 \le x \le L\}$$

where t is time and (x,r) are the axial and radial coordinates. The corresponding components of velocity will be denoted by (v_x, v_r) and pressure by p. For simplicity, the fluid flow is assumed to be axisymmetric. Note that the impermeability condition at the pipe walls leads to the equation

$$\frac{\partial a}{\partial t} + v_x \frac{\partial a}{\partial x} = v_r .$$

In particular, the radial component of velocity v_r at the pipe wall is expressed in terms of the given axial component $v_x = V(x,t)$ and the shape function $a(x,t)$.

Let the liquid be a Newtonian viscous fluid of dynamic viscosity μ and density ρ. The Navier-Stokes equations in the region D takes the form

$$\frac{\partial(r v_x)}{\partial x} + \frac{\partial(r v_r)}{\partial r} = 0, \tag{1}$$

$$\rho\left(\frac{\partial(v_x)}{\partial t} + v_x \frac{\partial v_x}{\partial x} + v_r \frac{\partial v_x}{\partial r}\right) = -\frac{\partial p}{\partial x} + \mu \Delta v_x , \tag{2}$$

$$\rho\left(\frac{\partial v_r}{\partial t} + v_x \frac{\partial v_r}{\partial x} + v_r \frac{\partial v_r}{\partial r}\right) = -\frac{\partial p}{\partial r} + \mu\left(\Delta v_r - \frac{v_r}{r^2}\right) , \tag{3}$$

with the boundary conditions on S

$$v_x = V, \quad v_r = \frac{\partial a}{\partial t} + V\frac{\partial a}{\partial x},$$

where

$$\Delta = \frac{\partial^2}{\partial x^2} + \frac{1}{r}\frac{\partial}{\partial r}\left(r\frac{\partial}{\partial r}\right) \tag{4}$$

is the Laplace operator.

3 Conservation laws

Let us derive the mass and momentum balance laws applied to a slice of the pipe. Integrating Eq. (1) with respect to r, using Eq. (4) and bearing in mind the regularity conditions of the solution on the axis of symmetry, the mass conservation law is obtained in the form

$$\frac{\partial A}{\partial t} + \frac{\partial Q}{\partial x} = 0 \tag{5}$$

where

$$A(x,t) = \pi a^2(x,t), \; Q(x,t) = 2\pi \int_0^{a(x,r)} v_x(x, r, t) r \, dr .$$

In the following, the mass conservation law will be taken in the form (5), whereas the momentum balance law will be reduced to an equation for Q and pressure p in order to construct an approximate model. Note that all the approximate models are based on the replacement of {a,r} with (εa, εr} where ε is a small parameter. Therefore, in order to make the mass balance law to be exact, the radial component of velocity v_r must be replaced with εv_r.
Analogous but more cumbersome calculations give the momentum conservation law in a slice as

$$\rho\left(\frac{\partial Q}{\partial t} + \frac{\partial W}{\partial x}\right) = F - \frac{\partial T}{\partial x} \tag{6}$$

where

$$W(x,t) = 2\pi \int_0^{a(x,t)} v_x^2(x,r,t)\, r\, dr\ ,$$

$$T(x,t) = 2\pi \int_0^{a(x,t)} v_x^2 \left[p(x,r,t) - 2\mu \frac{\partial v_x(x,r,t)}{\partial x} \right] r\, dr\ ,$$

and

$$F(x,t) = 2\pi a(x,t) \left[\left(p - 2\mu \frac{\partial v_x}{\partial x} \right)_{r=a(x,t)} \frac{\partial a(x,t)}{\partial x} + \mu \left(\frac{\partial v_x}{\partial r} + \frac{\partial v_r}{\partial x} \right)_{r=a(x,t)} \right].$$

The physical meaning of each term of these integral identities is obvious. In particular, A is the cross sectional area of the pipe and Q is the total flux of the fluid flowing through the pipe.

These conservation laws (5), (6) are precisely valid in a slice of the pipe, and can be used to test numerical algorithms, as well as to construct approximate equations based on postulating a certain *a priori* dependence of the solution with respect to the radial coordinate.

4. Lubrication theory

Formally, the lubrication theory approximation follows from the asymptotic expansion of the solution in a power series with respect to a small parameter ε, slenderness of the pipe. Since the continuity equation (1) and the adherence condition (4) are replaced with Eq. (5), one needs to simplify the momentum equations (2) and (3).

To avoid the procedure of non-dimensionalising the variables, let us formally stretch the variables, as in usually done in the lubrication theory. Namely, we replace $\{a, r, v_r, p\}$ by $\{\varepsilon a, \varepsilon r, \varepsilon v_r, \varepsilon^{-2} p\}$, and insert them into the governing Eqs. (2) and (3). As a result, the following system of equations is obtained to leading order [7],

$$\frac{\mu}{r} \frac{\partial}{\partial r} \left(r \frac{\partial v_x}{\partial r} \right) = \frac{\partial p}{\partial x},\ \frac{\partial p}{\partial r} = 0. \tag{7}$$

With the aid of Eqs. (4), Eqs. (7) lead to the parabolic profile of velocity (Hagen-Poiseuille flow),

$$v_x(x,r,t) = V(x,t) - \frac{a^2(x,t)-r^2}{4\mu}\frac{\partial p(x,t)}{\partial x}.$$

In particular,

$$Q(x,t) = \pi a^2(x,t)V(x,t) - \frac{\pi a^4(x,t)}{8\mu}\frac{\partial p(x,t)}{\partial x},$$

or equivalently

$$Q(x,t) = A(x,t)V(x,t) - \frac{A^2(x,t)}{8\pi\mu}\frac{\partial p(x,t)}{\partial x}, \tag{8}$$

The mass conservation law (5) which is not changed in the lubrication theory approach; in combination with Eq. (8) results in the equation for pressure

$$\frac{\partial}{\partial x}\left[\frac{A^2(x,t)}{8\pi\mu}\frac{\partial p(x,t)}{\partial x}\right] = \frac{\partial A(x,t)}{\partial t} + \frac{\partial[A(x,t)V(x,t)]}{\partial x},$$

$$p(0,t) = p_o(t),\ p(L,t) = p_1(t).$$

This simple system of equations is solved explicitly to give

$$\frac{\partial p(x,t)}{\partial x} = \frac{p_1(t) - p_0(t) + 8\pi\mu\int_0^L U(x,\xi,t)A^{-2}(\xi,t)d\xi}{A^2(x,t)\int_0^L A^{-2}(\xi,t)d\xi} \tag{9}$$

where

$$U(x,\xi,t) = \int_\xi^x \frac{\partial A(\eta,t)}{\partial t}d\eta + A(x,t)V(x,t) - A(\xi,t)V(\xi,t).$$

Note that $U(x,\xi,t) = U(\xi,x,t)$ which gives consistency of Eq. (9) with the boundary conditions for pressure. The total flux takes the form

$$Q(x,t) = \left[\int_0^L \frac{d\xi}{A^2(\xi,t)}\right]^{-1}\left\{\frac{p_0(t)-p_1(t)}{8\pi\mu} + \int_0^L\left[A(\xi,t)V(\xi,t) - \int_\xi^x \frac{\partial A(\eta,t)}{\partial t}d\eta\right]\frac{d\xi}{A^2(\xi,t)}\right\}. \tag{10}$$

Thus, all the flow characteristics are completely determined by the pressures drop $p_0(t) - p_1(t)$ along the pipe, and by the functions $A(x,t)$ and $V(x,t)$ describing the shape evolution of the pipe.

Some numerical calculations are shown in Figs. 1-5. The fluid viscosity is taken as $\mu = 0.1\, g/cm/\sec$, the pipe length is $L = 10\, cm$, the means radius of pipe is $R = 0.1\, cm$, the pressure drop is

$$\frac{p_0(t) - p_1(t)}{10^4\ N/cm^2} = 1 + 0.2\sin(2\pi\omega t)$$

where $\omega = 1/\sec$, and the cross-sectional area of the pipe is

$$\frac{A(x,t)}{\pi R^2} = 1 + \alpha \sin(2\pi\,\omega\, t)\sin(2\pi\kappa x / L)$$

where $\kappa = 2$ and $\alpha = 0.5$ (occulsion number).

5. Second-order approximation

Following the spirit of the theory of thin liquid films, a simple model accounting for inertial and viscous effects is here proposed. It is based on the following *a priori* radial dependence of velocity

$$v_x(x,r,t) = \frac{2Q(x,t)\left[A(x,t) - \pi r^2\right]}{A^2(x,t)}\ . \tag{11}$$

For simplicity, the case $V = 0$ is considered.

Stretching the variables in a slightly different manner from the lubrication theory approach, namely, replacing $\{a, r, v_r\}$ by $\{\varepsilon a, \varepsilon r, \varepsilon v_r\}$, let us insert them into Eqs. (2) and (3) and drop all the terms vanishing as $\varepsilon \to 0$. As a result, it is seen that Eq. (2) is unchanged, whereas Eq. (3) takes the form

$$\frac{\partial p}{\partial r} = \mu \frac{\partial}{\partial r}\left[\frac{1}{r}\frac{\partial (r\, v_r)}{\partial r}\right]\ .$$

In particular,

$$p(x,r,t) = P(x,t) + \frac{\mu}{r}\frac{\partial(r v_r)}{\partial r} \equiv P(x,t) - \mu\frac{\partial v_x}{\partial x}. \tag{12}$$

Inserting the expressions (11) and (12) into Eq. (2), multiplying by $2\pi r$ and integrating from 0 to $a(x,t)$, the second-order model obtained as

$$\frac{\partial A}{\partial t} + \frac{\partial Q}{\partial x} = 0, \tag{13}$$

$$\rho\left[\frac{\partial Q}{\partial t} + \frac{4}{3}\frac{\partial}{\partial x}\left(\frac{Q^2}{A}\right)\right] = 2\mu\left\{\frac{\partial^2 Q}{\partial x^2} - \frac{2Q}{A^2}\left[\left(\frac{\partial A}{\partial x}\right)^2 + 2\pi A\right]\right\} - A\frac{\partial P}{\partial x}. \tag{14}$$

Here $A(x,t)$ is given, whereas the modified pressure $P(x,t)$ and total flux $Q(x,t)$ are unknowns. The pressure P should be imposed at both ends of the pipe, and Q must be given at one side of the pipe, as well as initially.

Eqs. (13) and (14) account more precisely for inertial and viscous effects subject to the assumed parabolic profile. This approach is a modification of the well-known Galerkin method for solution of non-linear boundary value problems, which is based on an appropriate approximation of both the solution and governing equations of the problem by a finite set of basic functions. This method was applied to flow in a thin film [14, 15, 16, 17].

References

[1] Rath, H.J. (1980). *Peristaltische Stromungen*, Springer-Verlag, Berlin.

[2] Provost, A.M. and Schwartz, W.H. (1994). A theoretical study of viscous effects in peristaltic pumping, *J. Fluid Mech.*, **279**: 177-195.

[3] Pozrikidis, C. (1987). A study of peristaltic flow, *J. Fluid Mech.*, **180**: 515-527.

[4] Takabatake, S., Ayukawa, K. and Mori, A. (1988). Peristaltic pumping in circular cylindrical tubes: a numerical study of fluid transport and its efficiency. *J. Fluid Mech.*, **193**: 267-283.

[5] Brasseur, J. G. Corrsin, S. and Lu, N.Q (1987). The influence of peripheral layer of different viscosity on peristaltic pumping with Newtonian fluids. *J. Fluid Mech.*,**174**:495-519.

[6] Bohme, G. and Friedrich, R. (1983). Peristaltic flow of vicoelastic liquids. *J. Fluid Mech.*, **128**: 109-122.

[7] Shapiro, A.H., Jaffrin, M.Y. and Weinberg, S.L. (1969). Peristaltic pumping with long wavelengths at low Reynolds number. *J. Fluid Mech.*, **37**: 799-825.

[8] Brown, T.D. and Hung, T.K. (1977). Computational and experimental investigations of two-dimensional nonlinear peristaltic flows. *J. Fluid Mech.,* **83**:249-272.

[9] Chow, T.S. (1970). Peristaltic transport in a circular cylindrical pipe. *Trans. ASME J: J.Appl. Mech.*, **37**: 901-905.

[10] Fung, Y.C. and Yih, C.S. (1968). Peristaltic transport. *Trans. ASME J:J. Appl. Mech.*,**35**: 669-675.

[11] Shukla, J.B. and Gupta, S.P. (1982). Peristaltic transport of a power-law fluid with variable consistency. *Trans. ASME K: J. Biomech. Engng.*, **104**: 182-186.

[12] Jaffrin, M.Y. and Shapiro, A.H. (1971). Peristaltic pumping. *Ann. Rev. Fluid Mech.*, **3**: 13-36.

[13] Awon, M. P. Review of the literature with formulation of a theory for regulation of blood supply. (Private communication).

[14] Kapitsa, P. L. (1948). Wave flow of thin layers of viscous liquid. *Zh. Eksper. Teoret. Fiz.*, **18 (1)**: 3-28.

[15] Levich, V.G. (1967). *Physico-Chemical Hydrodynamics.* Prentice-Hall, Englewood Cliffs, NJ.

[16] Demekhin, E.A. and Shkadov, V. Ya. (1984). On three-dimensional unsteady waves in a moving liquid film. *Izv. AN* Nauk *SSSR Ser. Mekh. Zhidk. Gaza*, **5**: 21-27.

[17] Antanovskii, L.K. (1987). Generalized equations for a flow in a film on a plane wall. *Dinamika Sploshnoi Sredy*, **81**: 37-42. [In Russian].

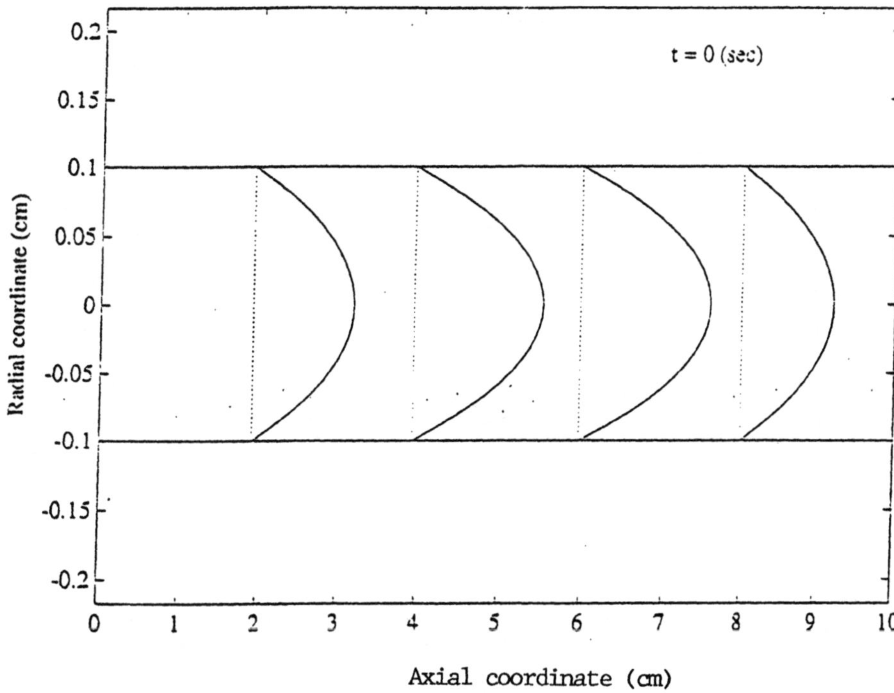

Figure 1: The shape of the pipe and velocity profiles

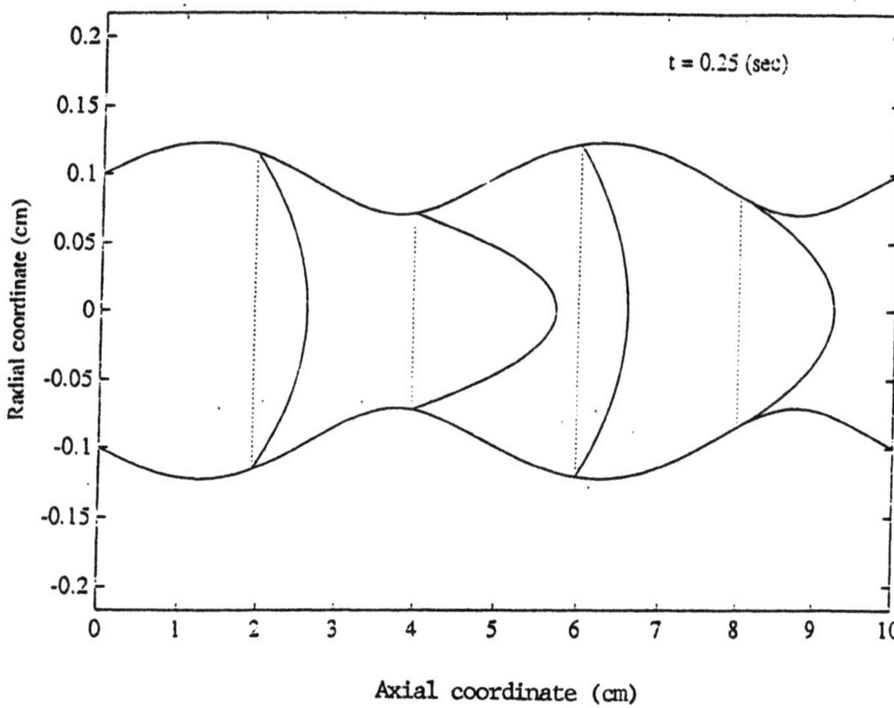

Figure 2: The shape of the pipe and velocity profiles

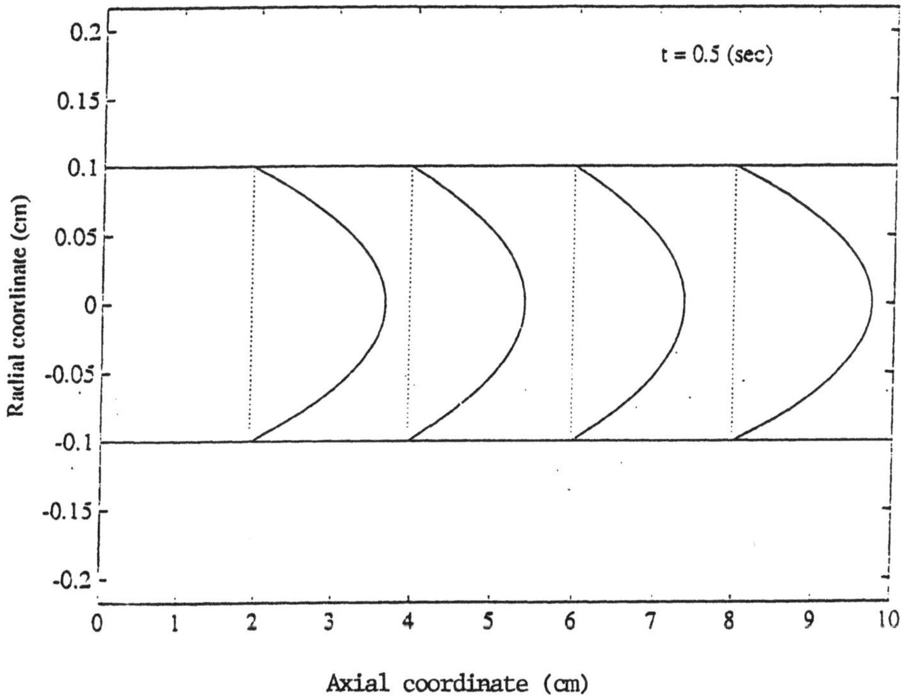

Figure 3: The shape of the pipe and velocity profiles

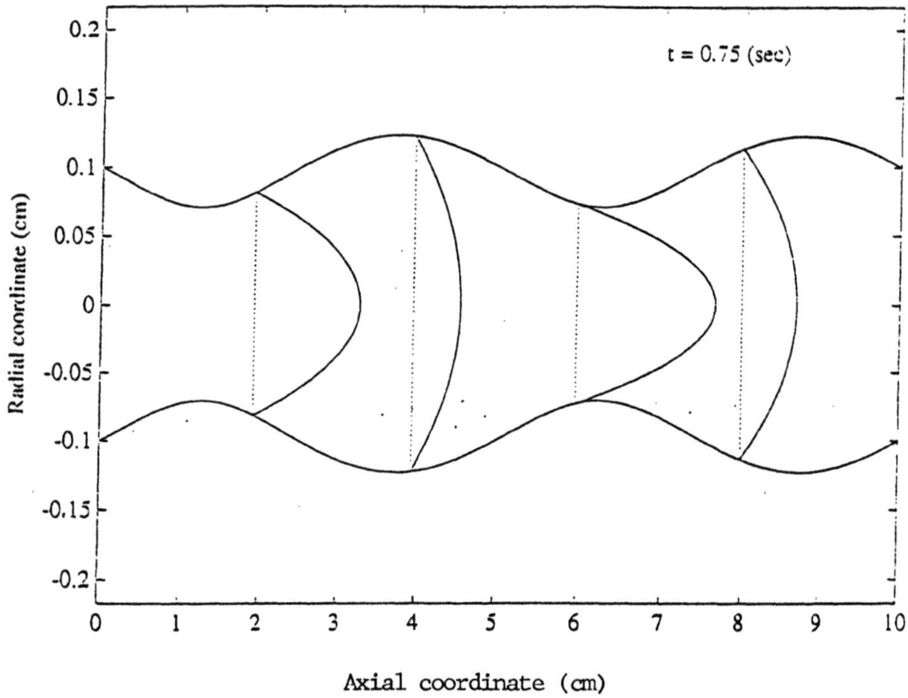

Figure 4: The shape of the pipe and velocity profiles

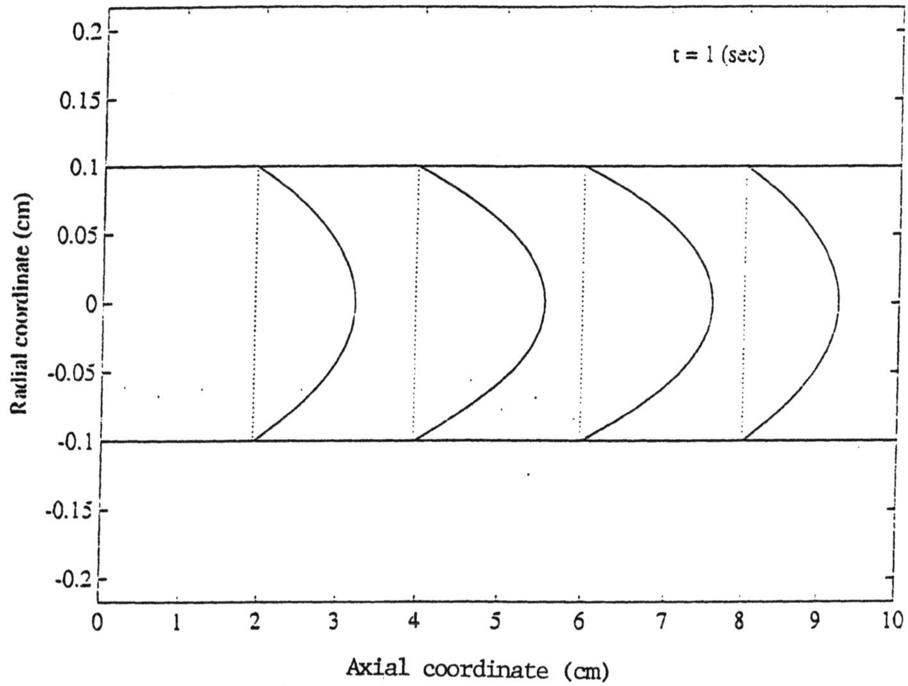

Figure 5: The shape of the pipe and velocity profiles

NUMERICAL SIMULATION OF PARTICLE MOTION IN VISCOELASTIC FLUIDS

Howard H. Hu
Department of Mechanical Engineering and Applied Mechanics
University of Pennsylvania
Philadelphia, PA 19104-6315

Abstract

A numerical method for computing motions of large numbers of particles (*particle mover*) in flows of solid-liquid (viscoelastic) mixtures was developed. In the method the fully coupled motions of liquid and solid are solved using a finite element technique, and solid particles move under the action of the hydrodynamic forces and moments exerted by the suspending fluid. The developed package is able to simulate the motion of particles in several popular models of non-Newtonian viscoelastic fluids.

This package was used to study the mechanisms of chaining of sedimenting particles in a viscoelastic fluid and detachment of the particles from the chain.

1. Introduction

Solid–liquid two phase flows are usually studied using a continuum approach that views solids and liquids as inter-penetrating mixtures, each being governed by conservation laws, either postulated or derived by averaging (see, Ishii 1975, Drew 1983, Joseph & Lundgren 1990, Zhang & Prosperetti 1994, for example). This approach results in unknown terms representing interactions between the liquid and solid phases. These terms must be modeled to close the equations (Anderson & Jackson 1967). The nature of the detailed interactions between solids and liquids cannot be understood from the application of the mixture theories.

The clusters and anisotropic microstructures observed in solid–liquid flows are the result of solid particle migrations produced by particle–particle and particle–wall interactions. These local rearrangement mechanisms are mediated by things like hydrodynamic forces and moments acting on the particles, wake interactions and vortex shedding. Direct numerical simulation of the exact particle motion in liquid may be the only theoretical tool capable of studying these nonlinear and geometrically complicated phenomena.

For simulations of solid–liquid flows at finite Reynolds numbers, Hu, Joseph & Crochet (1992) developed a numerical scheme using a finite element technique and simulated two-dimensional motions of a few (up to four) sedimenting circular and elliptic cylinders confined in a channel. This scheme solves the full nonlinear Navier-Stokes equation for the fluid in conjunction with the Newton's equation of motion for the solid rigid particles. Using the same scheme, Feng, Hu & Joseph (1994a, 1994b) studied the motion and interaction of circular and elliptical particles in sedimenting, Couette and Poiseuille flows of a Newtonian fluid. Huang, Feng & Joseph (1994) examined the turning couples on an elliptic particle settling in a channel. Hu (1995) studied the rotation of a circular cylinder settling close to a solid wall. Hu (1996) extended and improved the numerical scheme to compute the motion of hundreds circular particles in two-dimensions in a Newtonian fluid.

In this paper, a finite element technique based on moving unstructured grids to simulate motion of a large number of solid particles in viscoelastic fluids is described. When the suspending fluid is viscoelastic, we couple the fluid momentum equation with the constitutive equation for viscoelastic fluids. The system is solved with the EVSS (Elastic-Viscous-Split-Stress) formulation.

2. Description of the Problem

Consider an incompressible fluid occupying a bounded region Ω_t with boundary Γ_t at a given time instant *t*. There are *N* rigid solid objects (particles) freely moving in the fluid. We are interested in determining the motion of both fluid and individual solid particles. Since the Reynolds number of the flow based on particle size is not small, thus the inertia of the fluid and solid can not be neglected. The fluid motion has to satisfy the conservation of mass

$$\nabla \cdot \mathbf{u} = 0 \tag{1}$$

and the conservation of momentum

$$\rho_f\left(\frac{\partial \mathbf{u}}{\partial t} + (\mathbf{u}\cdot\nabla)\mathbf{u}\right) = \rho_f \mathbf{f} + \nabla\cdot\sigma \tag{2}$$

where $\mathbf{u}$ is the velocity vector, ρ_f is the density of the fluid, $\mathbf{f}$ is the body force, and σ is the stress tensor. For a Newtonian fluid the stress tensor is given by the simple constitutive relation

$$\sigma = -p\mathbf{I} + \eta\mathbf{D}[\mathbf{u}], \quad \mathbf{D}[\mathbf{u}] = \nabla\mathbf{u} + \nabla\mathbf{u}^T \tag{3}$$

where p is the pressure and μ_f is the viscosity of the fluid. For viscoelastic fluids, we choose the constitutive relations of the form

$$\sigma = -p\mathbf{I} + \eta_2\mathbf{D}[\mathbf{u}] + \tau_e \tag{4}$$

$$\text{where} \quad \lambda \overset{\nabla}{\tau}_e + z\tau_e + \alpha\frac{\lambda}{\eta_1}\tau_e\cdot\tau_e = \eta_1\mathbf{D}[\mathbf{u}] \tag{5}$$

$$\text{and} \quad \overset{\nabla}{\tau}_e = \frac{\partial \tau_e}{\partial t} + (\mathbf{u}\cdot\nabla)\tau_e - a\left(\nabla\mathbf{u}^T\cdot\tau_e + \tau_e\cdot\nabla\mathbf{u}\right) + (1-a)\left(\nabla\mathbf{u}\cdot\tau_e + \tau_e\cdot\nabla\mathbf{u}^T\right). \tag{6}$$

with proper choice of the parameters $(\lambda, \eta_1, \eta_2, z, \alpha, a)$ one can model fluids of Maxwell, Oldroyd B, Giesekus and PTT type.

The motion of solid objects satisfies the Euler's law

$$\frac{d}{dt}(\mathbf{MU}_p) = \mathbf{F}_p + \mathbf{G}_p \tag{7a}$$

and

$$\frac{d\mathbf{X}_p}{dt} = \mathbf{U}_p, \qquad \text{for} \quad p = 1,2,\ldots,N \tag{7b}$$

where $\mathbf{M}$ is the generalized mass matrix. $\mathbf{X}$ and $\mathbf{U}$ are, respectively, the generalized position and velocity vectors which incorporate both translational and angular components. The generalized force vector is expressed in two terms: $\mathbf{F}$ is the force imposed on the particle by the fluid and $\mathbf{G}$ is the body force exerted by external fields such as gravity. These matrix and vectors are written explicitly as

$$\mathbf{M} = \begin{bmatrix} m & 0 & 0 & 0 & 0 & 0 \\ 0 & m & 0 & 0 & 0 & 0 \\ 0 & 0 & m & 0 & 0 & 0 \\ 0 & 0 & 0 & I_x & I_{xy} & I_{xz} \\ 0 & 0 & 0 & I_{xy} & I_y & I_{yz} \\ 0 & 0 & 0 & I_{xz} & I_{yz} & I_z \end{bmatrix}, \ \mathbf{X} = \begin{bmatrix} X \\ Y \\ Z \\ \Theta_x \\ \Theta_y \\ \Theta_z \end{bmatrix}, \ \mathbf{U} = \begin{bmatrix} V_x \\ V_y \\ V_z \\ \Omega_x \\ \Omega_y \\ \Omega_z \end{bmatrix}, \ \mathbf{F} = \begin{bmatrix} F_x \\ F_y \\ F_z \\ T_x \\ T_y \\ T_z \end{bmatrix} \tag{8}$$

where m is the mass of the object, I_x, I_y, I_z are the moments of inertia and I_{xy}, I_{xz}, I_{yz} are the products of inertia of the object with respect to the (x,y,z) coordinates. For two-dimensional motions, the mass matrix reduces to diagonal $\mathbf{M} = \mathrm{diag}(m, m, I_z)$.

The forces and the moments imposed on an object by the fluid are given by

$$[F_x,\ F_y,\ F_z] = \int \sigma \cdot \hat{\mathbf{n}}\, d\Gamma \tag{9a}$$

and

$$[T_x,\ T_y,\ T_z] = \int (\mathbf{x} - \overline{\mathbf{X}}_p) \times (\sigma \cdot \hat{\mathbf{n}})\, d\Gamma \tag{9b}$$

where the integration is performed over the surface of the object. $\hat{\mathbf{n}}$ is the unit normal vector on the surface of the object pointing outward. $\overline{\mathbf{X}}_p = [X, Y, Z]^T$ is the position of the centroid of the object.

The boundary of the domain occupied by the fluid, Γ_t, can be decomposed into three non-overlapping sections: $(\Gamma_t)_u$, $(\Gamma_t)_\sigma$ and $\bigcup_p (\Gamma_t)_p$, where $(\Gamma_t)_p$ is the surface of solid particle p. On these boundary sections the following types of boundary conditions are imposed

$$\mathbf{u} = \mathbf{g}, \text{ on } (\Gamma_t)_u \tag{10.a}$$

$$\sigma \cdot \mathbf{n} = \mathbf{h}, \text{ on } (\Gamma_t)_\sigma \tag{10.b}$$

and
$$\mathbf{u} = \mathbf{V}_p + \Omega_p \times (\mathbf{x} - \overline{\mathbf{X}}_p), \text{ on } (\Gamma_t)_p, \quad p = 1,2,\ldots,N \tag{10.c}$$

where $\mathbf{g}$ and $\mathbf{h}$ are prescribed values of velocity and traction. $\mathbf{n}$ is the unit normal vector on the boundary. $\mathbf{V}_p$ and Ω_p are, respectively, the transitional and angular velocities of object p $\left(\mathbf{U}_p = \left[\mathbf{V}_p, \Omega_p\right]^T\right)$. (9) represents the no-slip condition on the surface of the object.

3. Galerkin Finite Element Formulation

We seek a generalized Galerkin finite element formulation that incorporates both the fluid and particle equations of motion. In the formulation, the spatial coordinates are discretized using finite element scheme, and the temporal coordinate is treated with a finite difference method.

At time instant t, the fluid occupies the domain Ω_t with boundary Γ_t. We define the finite element interpolation functional space for the velocity and the pressure in the fluid phase as

$$S_{\mathbf{u}}^h = \left\{ \mathbf{u}^h = \left(u_x^h, u_y^h, u_z^h\right) \mid \mathbf{u}^h \in H^{1h}(\Omega_t),\ \mathbf{u}^h = \mathbf{g}^h \text{ on } (\Gamma_t)_u, \right.$$
$$\left. \text{and } \mathbf{u}^h = \mathbf{V}_p + \Omega_p \times \left(\mathbf{x} - \overline{\mathbf{X}}_p\right) \text{ on } (\Gamma_t)_p, \forall p = 1,\ldots,N \right\},$$

$$V_{\mathbf{w}}^h = \left\{ \mathbf{w}^h = \left(w_x^h, w_y^h, w_z^h\right) \mid \mathbf{w}^h \in H^{1h}(\Omega_t),\ \mathbf{w}^h = 0 \text{ on } (\Gamma_t)_u, \right.$$
$$\left. \text{and } \mathbf{w}^h = \delta\mathbf{V}_p + \delta\Omega_p \times \left(\mathbf{x} - \overline{\mathbf{X}}_p\right) \text{ on } (\Gamma_t)_p, \forall p = 1,\ldots,N \right\}$$

and
$$S_p^h = V_p^h = \left\{ p^h \mid p^h \in H^{1h}(\Omega_t) \right\}$$

where $\mathbf{V}_p$ and Ω_p $\left(\mathbf{U}_p = \left[\mathbf{V}_p, \Omega_p\right]^T\right)$ are the translational and angular velocities of the solid particles, and $\delta\mathbf{V}_p$ and $\delta\Omega_p$ $\left(\delta\mathbf{U}_p = \left[\delta\mathbf{V}_p, \delta\Omega_p\right]^T\right)$ are their variations (arbitrary functions of time). In the finite element domain, these spaces are formed using second-order polynomials for the velocity and first-order polynomials for the pressure (P_2/P_1 elements).

An EVSS (Elaastic-viscous-split-stress) Galerkin finite element formulation can be expressed as:

Find $\mathbf{u}^h \in S_{\mathbf{u}}^h$, $p^h \in S_p^h$ and τ^h and $\tau_{\mathrm{N}}{}^h \in S_p^h$ and such that $\forall\, \mathbf{w}_{\mathrm{u}}{}^h \in V_{\mathbf{w}}^h$, $q^h, \mathbf{w}_\tau{}^h$ and $\mathbf{w}_{\mathrm{D}}{}^h \in V_p^h$

$$\begin{aligned}
&\int_\Omega \mathbf{w}_{\mathrm{u}} \cdot \rho\left[\frac{\delta\mathbf{u}}{\delta t} + (\mathbf{u}-\mathbf{u}_m)\cdot\nabla\mathbf{u} - \mathbf{f}\right] d\Omega + \int_\Omega \nabla\mathbf{w}_{\mathrm{u}} : \left[\eta\left(\nabla\mathbf{u}+\nabla\mathbf{u}^T\right)+\tau\right] d\Omega \\
&\qquad -\int_\Omega (\nabla\cdot\mathbf{w}_{\mathrm{u}})\, p\, d\Omega + \int_\Omega w_{\mathrm{p}} \nabla\cdot\mathbf{u}\, d\Omega \\
&\qquad +\int_\Omega \mathbf{w}_{\mathrm{D}} : \left[\eta_1\left(\nabla\mathbf{u}+\nabla\mathbf{u}^T\right)\right] d\Omega \\
&+\int_\Omega \mathbf{w}_\tau : \left[\lambda\,(\tau \overset{\nabla}{+} \tau_N) + z(\tau+\tau_N) + \alpha\frac{\lambda}{\eta_1}(\tau+\tau_N)\cdot(\tau+\tau_N) = \tau_N\right] d\Omega \\
&\qquad \sum_p \mathbf{W}_{\mathrm{p}} \cdot \left(\mathbf{M}\frac{d\mathbf{U}_p}{dt} - \mathbf{G}_p\right) = 0
\end{aligned} \tag{11}$$

where $\tau_N = \eta_1\left(\nabla\mathbf{u}+\nabla\mathbf{u}^T\right)$. To handle the movement of the finite element mesh, an arbitrary Lagrangian-Eulerian (ALE) technique is used. In (11), $\delta\mathbf{u}^h/\delta t$ is the time derivative following the same referential mesh node. $\mathbf{u}_m^h$ is the velocity vector which defines the motion of mesh nodes. In general, the mesh velocity can be specified arbitrarily as long as it follows the motion of the particles (slip on the particle surface is also allowed). In our implementation, we assume that the nodes on the particle surface move with the particle (no slip), and the nodes in the interior of the fluid are computed using Laplace's equation, to guarantee a smoothly varying distribution of the nodes:

$$\nabla^2\mathbf{u}_m = 0 \quad \text{in } \Omega_t, \tag{12}$$

with boundary conditions given by

$$\mathbf{u}_m = \mathbf{u} = \mathbf{V}_p + \Omega_p \times \left(\mathbf{x}-\overline{\mathbf{X}}_p\right) \quad \text{on } (\Gamma_t)_p, \quad p = 1,2,\ldots,N$$

and

$$\mathbf{u}_m = 0 \quad \text{on } (\Gamma_t)_u \cup (\Gamma_t)_\sigma. \tag{13}$$

Equation (13) is also solved with a Galerkin finite element formulation whose functional space is formed using first-order (linear) polynomials, to reduce computational cost. The values of the mesh velocity on the remaining mesh nodes are interpolated locally inside each finite element.

4. Solution Procedure

Initially, the particles are positioned randomly in the fluid, with zero velocity. The fluid is either at rest or flowing steadily around the particles. The particles are then released, and the motion of the combined fluid/particle system is simulated. Hu, et al. (1992) developed a solution procedure in which the positions of the particles and the grid nodes in the fluid domain are updated explicitly, while the particle velocities and the fluid flow field (velocity and pressure) are determined implicitly to avoid numerical instabilities. A modified version of that procedure which incorporates the ALE scheme with mesh velocities determined implicitly by solving a set of Laplace's equations is implemented in this study. The details of the modified solution procedure are listed below.

(1) Initialization.

$t_0 = 0,\ n = 0$ (index for time step).

Generate an initial mesh $\mathbf{x}_0$.

Initialize $\mathbf{u}(\mathbf{x}_0, 0)$, $p(\mathbf{x}_0, 0)$ and $\mathbf{X}_p(0)$, $\mathbf{U}_p(0)$ for $p = 1, 2, \ldots, N$;

(2) Explicit Update.

Select time step Δt_{n+1}: $t_{n+1} = t_n + \Delta t_{n+1}$.

Update particle position $\mathbf{X}_p(t_{n+1}) = \mathbf{X}_p(t_n) + \Delta t_{n+1} \mathbf{U}_p(t_n)$, for $p = 1, 2, \ldots, N$.

Update mesh nodes $\hat{\mathbf{x}}(t_{n+1}) = \mathbf{x}(t_n) + \Delta t_{n+1} \mathbf{u}_m(t_n)$.

(3) Remeshing and Projection (if the mesh $\hat{\mathbf{x}}(t_{n+1})$ is too distorted).

Generate a new mesh $\mathbf{x}(t_{n+1})$.

Project the flow field from $\hat{\mathbf{x}}(t_{n+1})$ onto $\mathbf{x}(t_{n+1})$.

(4) Flow Solver.

Iteratively solve $\mathbf{u}(\mathbf{x}(t_{n+1}), t_{n+1})$, $p(\mathbf{x}(t_{n+1}), t_{n+1})$, $\mathbf{u}_m(\mathbf{x}(t_{n+1}), t_{n+1})$ and $\mathbf{U}_p(t_{n+1})$ for $p = 1, 2, \ldots, N$.

(5) If the time t_{n+1} is less than a specified time go to step (2); otherwise stop.

In the procedure, the size of the time step Δt_{n+1} is determined by the maximum distance each particle is allowed to travel, or the maximum change in the particle speed. The explicit update step (2) is first order accurate in time, which can be easily improved to second order by introducing accelerations of the solid particles and the mesh nodes. In the flow solver, the system of equations is solved with an iterative solver using a preconditioned generalized minimal residual (GMRES) procedure. The size of the Krylov space is set to 20. We also implemented a preconditioner using incomplete LU factorization with controlled threshold fill-in elements (ILUT). The number of fill-ins is controlled to be less than 10 for both the upper and lower triangular matrices.

5. Chain Snapping

When a chain of spheres falling behind a single sphere in a Newtonian or viscoelastic fluid at very low Reynolds numbers, they converge and form a longer chain (see Fig. 1). This is because a long chain of spheres behaves like a long object and thus tends to fall faster than a single sphere. A long body falls faster since its effective weight increases in proportion to its length while the drag exerted by the fluid on it does not increase in the same proportion. This causes a kissing contact between the chain of spheres and the single sphere. Due to inertial effects this long chain structure becomes unstable and tumbles at higher Reynolds number. Such turning happens for finite Reynolds number in Newtonian fluids, whereas in viscoelastic fluids this happens at relatively higher Reynolds number depending on the elastic properties of the fluids (Liu and Joseph [9]).

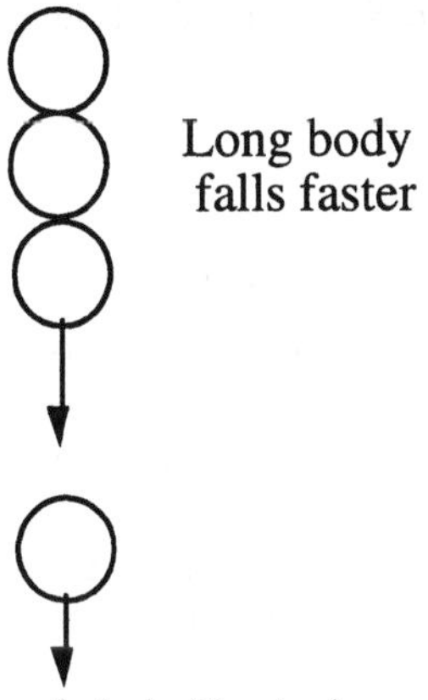

Fig. 1. Formation of a long chain in Newtonian and Viscoelastic fluids. This chain configuration is unstable in Newtonian fluids whereas it is stable in viscoelastic fluids in slow flows.

In the case of a single sphere falling behind a chain of spheres, there are various mechanisms at work. The trailing sphere will not always be attracted to the chain to form a longer chain. In viscoelastic fluids, it is seen that there will be attraction if the separation between the sphere and the chain is not very large. As depicted in Fig. 2, the trailing sphere will separate from the chain when the distance between the last two spheres exceeds a critical value (Joseph,1996). However, sometimes attached sphere in a chain gets separated. We call this chain snapping. It is difficult to experimentally study this phenomenon in Newtonian fluids since the perfectly aligned chain configuration is not stable. However, in numerical simulations one can overcome this difficulty.

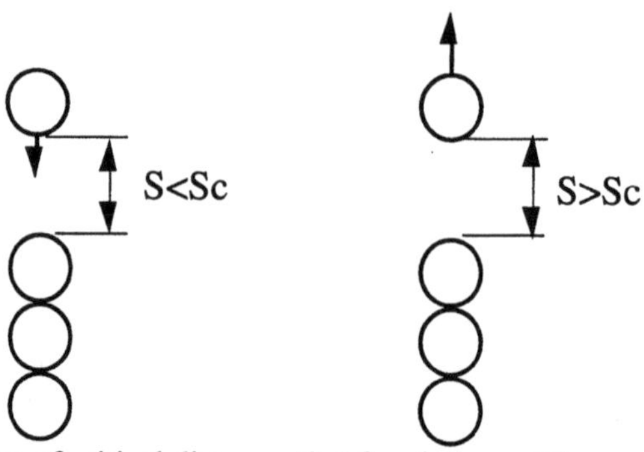

Fig. 2. Phenomenon of critical distance (Sc) for chains settling in viscoelastic fluids.

In this section we address the mechanism of chain snapping. We perform two-dimensional dynamic simulations of particles in Newtonian and Oldroyd-B fluids. In our simulation, particles are arranged such that there are N particles (two dimensional cylinders) in the leading chain and one particle is placed behind the leading chain with some initial separation. We found that chain snapping is not peculiar to viscoelastic fluids. It can also be observed in Newtonian fluids under certain ideal conditions.

5.1 Nondimensional Parameters

Consider a viscoelastic (Oldroyd-B) fluid with constant properties. Let η be the viscosity, λ_1 be the relaxation time and λ_2 be the retardation time of the fluid. Let d be the diameter of the cylinder, g be the gravitational acceleration, ρ_s be the density of the particle and ρ_f be the density of the fluid. Let S_c be the critical distance of separation in which we are primarily interested and let there be N particles in the leading chain. Assume that the width of the channel is large enough to exclude the wall effects. In general the critical distance of separation will be a function of all variables involved:

$$S_c = f(d, \rho_s, \rho_f, \eta, \lambda_1, \lambda_2, g, N), \tag{14}$$

Applying the Buckimgham's Pi theorem we can write,

$$\left(\frac{S_c}{d}\right) = f\left(N, \frac{\rho_s}{\rho_f}, \frac{\eta}{\rho_f d\sqrt{gd}}, \frac{\lambda_1 \eta}{\rho_f d^2}, \frac{\lambda_2}{\lambda_1}\right). \tag{15}$$

Note that

$$\frac{\eta}{\rho_f d\sqrt{gd}} \equiv \frac{Fr}{\text{Re}} \quad \text{and} \quad \frac{\lambda_1 \eta}{\rho_f d^2} \equiv \frac{De}{\text{Re}},$$

where Re is the Reynolds number, Fr is the Froude number and De is Deborah number of the flow. Since the characteristic velocity of the problem is not known beforehand we consider the ratio of the dimensionless numbers in the expression for critical distance. In the numerical simulation we use an Oldroyd-B model with $\lambda_2/\lambda_1 = 1/8$. Note that $\lambda_1 = 0$ gives Newtonian fluid as a special case of the above.

5.2 Chaining in a Newtonian Fluid

A vertically falling chain is unstable in Newtonian fluids. In the simulations we consider a chain sedimenting along the center line of a vertical channel (Fig. 3). The width of the channel is taken large enough to exclude the wall effects. Due to the symmetry of the problem the vertical chain configuration is stable during the simulations even in a Newtonian fluid. We shall now demonstrate how the critical distance varies with these various parameters.

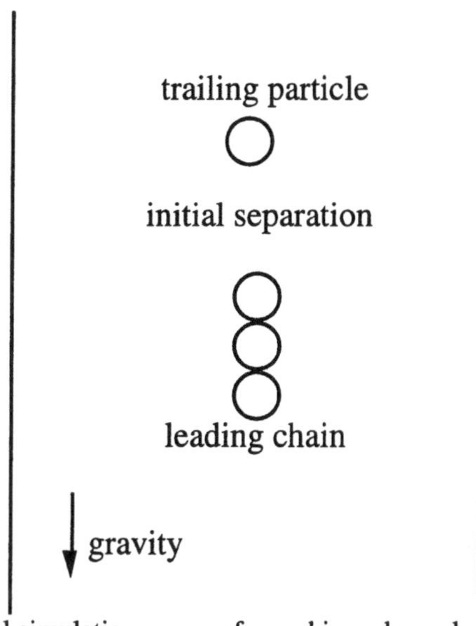

Fig. 3. Numerical simulations are performed in a channel with walls wide apart (channel width is 15 times the particle diameter).

5.3 Phenomenon of critical distance of separation

To demonstrate the phenomenon of critical distance of separation in a Newtonian fluid, we consider a case with two particles in the chain, $N = 2$, $\rho_s/\rho_f = 1.01$ and $Fr/\mathrm{Re} = 0.026$. The result is presented in Fig. 4. There we plot the velocities of the leading chain and the trailing particle as a function of time for three different initial separations between the leading chain and the trailing particle. If the trailing particle falls with a velocity lower than that of the leading chain separation occurs, whereas if it falls faster attachment occurs. Fig. 4 shows that for zero initial separation all the particles fall with a common velocity indicating no detachment. When the initial separation is increased to 1.5 times the diameter of the particles we see the trailing particle falls faster indicating attachment. On further increasing the initial distance of separation to 2 times the diameter of the particle we see that the trailing particle now falls slower than the leading chain indicating further separation. This clearly indicates that the phenomenon of critical distance in sedimenting chain is not peculiar to the viscoelastic fluid model and can be seen in Newtonian fluids as well.

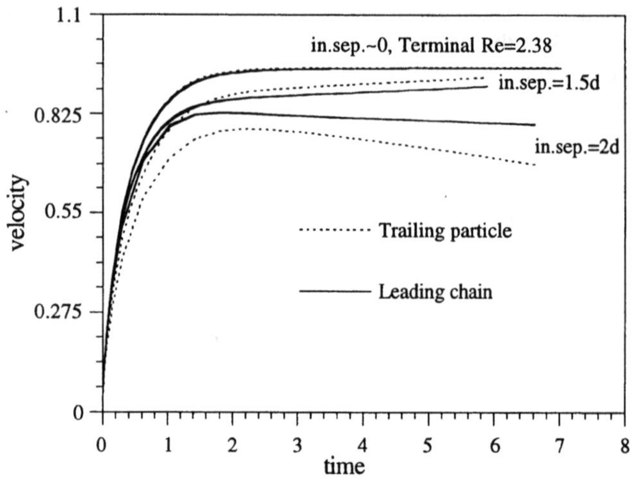

Fig. 4. Phenomenon of critical distance in Newtonian fluids, $N = 2$, $\rho_s / \rho_f = 1.01$ and $Fr / \mathrm{Re} = 0.026$.

This phenomenon can be easily explained. The leading chain behaves like a long body and tends to fall faster than a single particle. Let us call this the long body effect. Consequently if the trailing particle is far behind the leading chain the distance between them increases. However, since the trailing cylinder is in the wake of the leading chain it experiences reduced drag and may fall faster than the leading thus attachment may occur. Therefore we see attraction due to the wake effect. Competition between the long body effect and the wake effect gives rise to the phenomenon of critical distance of separation in a Newtonian fluid. Each of these two effects depends on parameters like the number of particles in the leading chain, density ratio and viscosity of the fluid.

5.4 Effect of the number of particles in the leading chain

We now consider a case with $\rho_s / \rho_f = 1.005$ and $Fr / \mathrm{Re} = 0.026$. The initial distance of separation between the leading chain and the trailing particle is zero. Thus there are a total number of $N+1$ particles in the sedimenting chain. We shall consider two cases with $N = 2$ and 3. Fig. 5 shows that for $N = 2$ all the particles fall with a common velocity indicating no detachment. For $N = 3$ the leading chain falls faster than the trailing particle indicating detachment. Thus we see that increasing the number of particles in the leading chain induces detachment, and there is a maximum number of particles in a sedimenting chain.

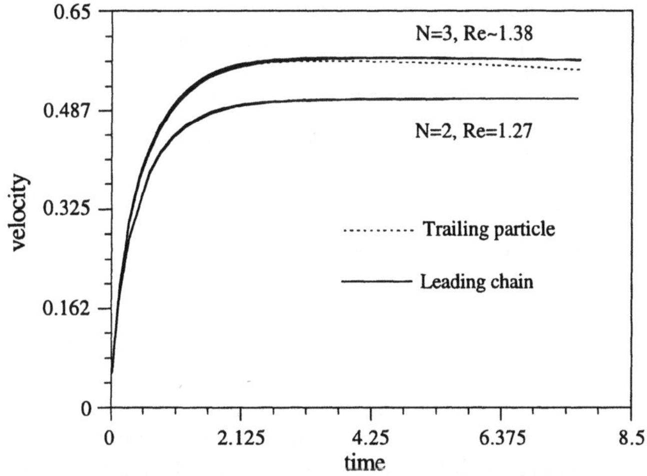

Fig. 5. Effect of number of particles in the leading chain, initial separation~0, $\rho_s / \rho_f = 1.005$ and $Fr / \mathrm{Re} = 0.026$.

This can be explained based on the mechanisms proposed above, namely the long body effect and the wake effect. Increasing the number of particles in the leading chain increases both, the long body effect (since the leading chain becomes longer) and the wake effect (since there will be a stronger wake behind a faster falling chain). Above observation shows that the long body effect dominates inducing detachment when the number of particles in the leading chain is increased. This causes the critical distance of separation to decrease with the increase in the number of particles in the leading chain.

5.5 Effect of viscosity

Consider the case with $N = 3$ and $\rho_s / \rho_f = 1.01$. The initial distance of separation between the trailing particle and the leading chain is zero. The increase in viscosity is manifested through the increase in the ratio of Froude and Reynolds number. We shall consider two cases with $Fr / \mathrm{Re} = 0.026$ and 0.052. Fig. 6 shows that the trailing particle remains attached to the leading chain when $Fr / \mathrm{Re} = 0.026$. Increase in the ratio of Froude and Reynolds number causes detachment of the trailing cylinder. This means that increasing the viscosity induces detachment.

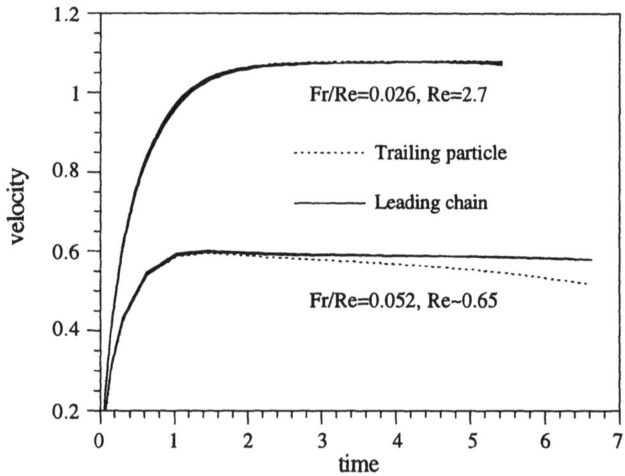

Fig. 6. Effect of viscosity, initial separation = 0, $N = 3$ and $\rho_s / \rho_f = 1.01$.

Increase in viscosity (or increase in the ratio of Froude and Reynolds number) increase the drag on the particles and reduces the Reynolds number (see Fig. 6). Inertia effects become less dominant at lower Reynolds number reducing the wake effect. Since the wake effects become weak with increase in the viscosity we see detachment. This causes the critical distance of separation to decrease with an increase in the ratio of the Froude and Reynolds number.

Effects of these various parameters on the phenomenon of detachment corroborate the mechanism of attraction (wake effect) and the mechanism of separation (long body effect) in chains. Another parameter, namely the fluid elasticity will play a role in chaining when we consider viscoelastic fluids.

5.6 Chaining in Viscoelastic fluids

We consider next an Oldroyd-B fluid with $\lambda_2 / \lambda_1 = 1/8$. The ratio of Deborah and Reynolds number will be non-zero. We can study the effect of viscoelastic properties by directly comparing the phenomenon of chaining in the Oldroyd-B fluid with that of the Newtonian fluid keeping all the other parameters same. To accomplish such a comparison consider a case with $N = 2$, $\rho_s / \rho_f = 1.003$ and $Fr / \mathrm{Re} = 0.026$. The initial separation between the trailing particle and the leading chain is zero. We shall compare the results for a Newtonian fluid $De / \mathrm{Re} = 0$ with that of an Oldroyd-B fluid with $De / \mathrm{Re} = 2.4$. Fig. 7 shows that the trailing particle gets detached in a Newtonian fluid whereas in an Oldroyd-B fluid there is no detachment. This suggests that elasticity of the fluid further induced attachment.

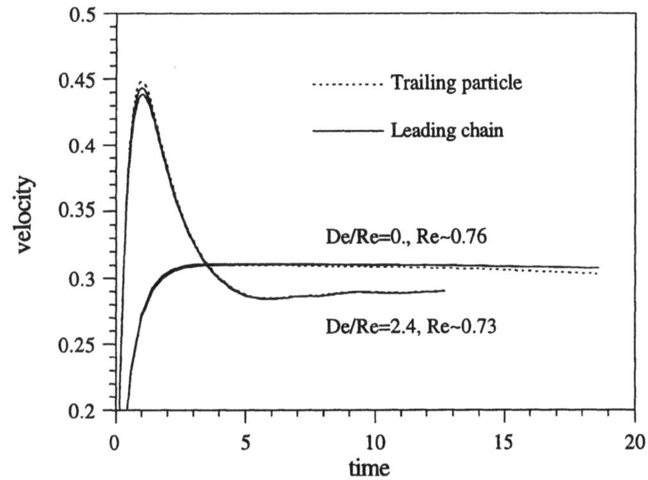

Fig. 7. Effect of fluid elasticity, initial separation~0, $N=2$, $\rho_s/\rho_f=1.003$ and $Fr/\mathrm{Re}=0.026$.

This tendency to attach in a viscoelastic fluid is due to the normal stress effects as explained by Joseph (1996). It was pointed out that one only need to examine the crowded streamlines in the flow near the body to see how the normal stresses are distributed over the body. Two particles settling in tandem experience imbalanced compressive normal stresses at the bottom of the leading particles and the top of the trailing particle, causing them to chain. Thus in viscoelastic fluids there is one more mechanism of attraction, namely the normal stress effect.

Fig. 8 shows the phenomenon of chain snapping in a viscoelastic fluid. It shows that the last two particles in a chain of six particles get detached from the chain one after the other.

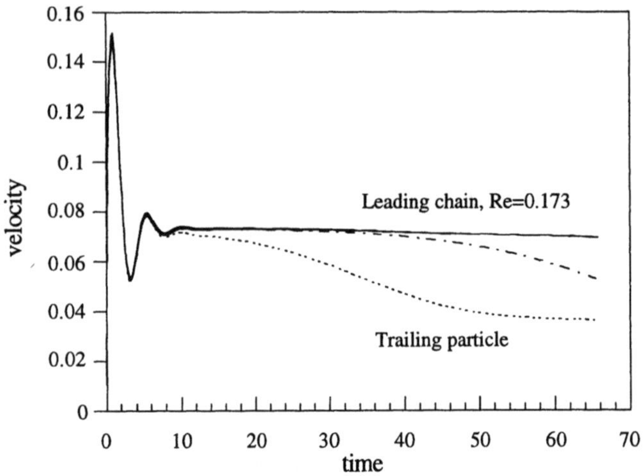

Fig. 8. Sequential chain snapping of a chain of six particles in a viscoelastic fluid, $\rho_s / \rho_f = 1.001$, $Fr / \mathrm{Re} = 0.026$ and $De / \mathrm{Re} = 2.4$.

6. Conclusions

In this paper, a numerical method for computing motions of large numbers of particles (*particle mover*) in flows of solid-liquid mixtures was described. In the method the fully coupled motions of liquid and solid are solved using a finite element technique, and solid particles move under the action of the hydrodynamic forces and moments exerted by the suspending fluid. The developed scheme is able to simulate the motion of particles in several popular models of non-Newtonian viscoelastic fluids.

Using this numerical scheme, we investigated the phenomenon of detachment of a trailing particle from a leading chain of particles frequently observed in sedimentation experiments involving viscoelastic fluids. We carried out two-dimensional dynamic simulations of particles in Newtonian and Oldroyd-B fluids. We found that chain snapping is not peculiar to viscoelastic fluids. It can also be observed in Newtonian fluids under certain ideal conditions.

In the case of a single particle falling behind a chain of particles in a Newtonian fluid there are various mechanisms at work. The chain of particles in front behaves like a long body. Hence it tends to fall faster than a single sedimenting particle, causing the separation of the leading chain and the trailing particle. This is the long body effect. The mechanism that works against this is the wake effect. The wake of the leading chain causes drag reduction on the trailing particle, which induces the attraction. This attraction comes into effect only when the inertial effects are significant. At lower Reynolds numbers the wake effect is weak causing the trailing particle to separate even for zero initial separation from the leading chain.

When a particle falls behind a chain of particles in a viscoelastic fluid, in addition to the above two effects, namely the long body effect and the wake effect, there is a normal stress effect in action. Normal stress effects cause attraction between the trailing sphere and the leading chain. This effect works against the long body effect.

In summary, long body effect causes detachment. Normal stress and wake effects cause attraction of the trailing particle. They compete with one another giving rise to the phenomenon of critical distance of separation. If the initial separation of the trailing particle from the leading chain is less than this critical distance there will be attraction otherwise there will be separation. The critical distance of separation depends on the various flow parameters. If the functional form of the dependence is known then it can be used to determine some rheological property of the fluid.

Acknowledgment
This work is supported by National Science Foundation HPCC Grand Challenge Grant ESC-95-27123, National Science Foundation grant CTS 94-10022 and by the Research Foundation of the University of Pennsylvania.

References

Anderson, T. B. & Jackson, R.(1967) A fluid mechanical description of fluidized beds: Equations of motion. *Ind. Engng. Chem. Fundam.* **6**, 527-539.

Drew, D.(1983) Mathematical modeling of two-phase flow. *Ann. Rev. Fluid Mech.* **15**, 261-291.

Feng, J., Hu, H.H. & Joseph, D.D.,(1994a,) Direct simulation of initial value problems for the motion of solid bodies in a Newtonian fluid. Part 1: sedimentation, *J. Fluid Mech.* **261**, 95-134.

Feng, J., Hu, H.H. & Joseph, D.D.,(1994b) Direct simulation of initial value problems for the motion of solid bodies in a Newtonian fluid. Part 2: Couette and Poiseuille flows *J. Fluid Mech.* **277**, 271-301.

Hu, H.H., "Motion of a circular cylinder in a viscous liquid between parallel plates", *Theoretical and Computational Fluid Dynamics.* (1995), in print.

Hu, H. H., Joseph, D. D. & Crochet, M. J.(1992) Direct simulation of fluid particle motions. *Theoret. & Comput. Fluid Dyn.* **3**, 285-306.

Hu, H.H.,(1996,) Direct simulation of flows of solid-liquid mixtures, *Int. J. Multiphase Flow* **22**, 335-352.

Huang, P.Y., Feng, J. & Joseph, D.D.(1994,)The turning couples on an elliptic particle settling in a vertical channel, *J. Fluid Mech.* **271**, 1-16.

Ishii, M.(1975) *Thermo-Fluid Dynamic Theory of Two-Phase Flows.* Eyrolles, Paris.

Joseph, D.D. & Lundgren, T.S.,(1990,) Ensemble averaged and mixture theory equations for incompressible fluid particle suspensions. *Int. J. Multiphase Flow* **16**, 35-42.

Joseph, D.D., (1996,)Flow induced microstructure in Newtonian and Viscoelastic fluids, Keynote presentation at the 5th World Congress of Chemical Engineering, Particle Technology Track, Second Particle Technology Forum, San Diego.

Liu, Y.J. and D.D. Joseph, (1993,) Sedimentation of particles in polymer solutions, *J. Fluid Mech.* **255**, 565-595.

Zhang, D.Z., & Prosperetti, A., (1994,)Averaged equations for inviscid disperse two-phase flow, *J. Fluid Mech.* **267**, 185-219.

MIGRATIONAL INSTABILITY IN SHEAR-THINNING SUSPENSIONS

J.D. GODDARD
Department of Applied Mechanics and Engineering Sciences
University of California, San Diego
La Jolla, CA 92093-0411, U.S.A.

Abstract.

This work deals with an instability arising from the shear-induced migration of particles in dense suspensions coupled with a dependence of viscosity on particle concentration. As an extension of a recent work by the present author treating the inertialess ($Re = 0$) linear stability of homogeneous simple shear flow, the current article includes shear-thinning effects. Depending on the importance of shear-induced migration relative to concentration-driven diffusion, the short-wave instability found in the previous work is shown to be enhanced by shear thinning. Such instability would lead ultimately to a two-phase structure, involving particle-rich regions in dynamic equilibrium with particle-depleted regions, which in the case of pipe flow should result in a "core-annular" structure, with viscous core surrounded by a less viscous lubricating layer.

1. Introduction

This is an extension of a previous analysis [1] of migrational instabilities in sheared particle suspensions, whose goal is to explore the influence of certain non-Newtonian effects on stability. The previous work employed as constitutive prototype the neutrally-buoyant "Stokesian" suspension model of Leighton and Acrivos [2] (*cf.* [3] and [4]), in which the stress and the particle flux relative to the mixture are given, respectively, by

$$\mathbf{T} = 2\eta(\phi)\mathbf{D} - p\mathbf{1} \tag{1}$$

and

$$\mathbf{j} = -\{\kappa(\phi, \dot{\gamma})\nabla\phi + \nu(\phi, \dot{\gamma})\nabla\dot{\gamma}\} \tag{2}$$

where ϕ denotes particle volume fraction,

$$\mathbf{D} := \frac{1}{2}\{\nabla\mathbf{v} + (\nabla\mathbf{v})^T\}, \quad \dot{\gamma} := \left\{2\mathrm{tr}\left(\mathbf{D}^2\right)\right\}^{\frac{1}{2}} \tag{3}$$

and $\mathbf{v}$ is mixture velocity. This model, or certain variants thereof, have been put forth to explain particle migration effects in suspensions of rigid particles. In the case of deformable particles such as emulsion droplets, blood cells, or large globular macromolecules, etc., one may expect to find similar, possibly much stronger migration effects [5] accompanied also by non-Newtonian effects such as shear thinning. After a brief recapitulation of the previous analysis, we set down the modification arising from the presence of shear thinning.

2. Previous Analysis

The quasi-static mechanics is governed by the above constitutive equations, the balances

$$\nabla \cdot \mathbf{T} = \nabla p \ \text{ with } \ \nabla \cdot \mathbf{v} = 0, \tag{4}$$

and

$$\partial_t \phi + \mathbf{v} \cdot \nabla\phi + \nabla \cdot \mathbf{j} = 0, \tag{5}$$

together with suitable boundary conditions (compatibility with the base state and/or regularity of perturbed states in unbounded domains).

For planar infinitesimal perturbations $\mathbf{v}^{(1)}, \phi^{(1)}$ of the uniform base state:

$$v_x^{(0)} = \dot{\gamma}^{(0)} y \ \text{ and } \ \{v_y^{(0)}, v_z^{(0)}, \nabla\dot{\gamma}^{(0)}, \nabla\phi^{(0)}\} = 0, \tag{6}$$

the Fourier representation of the perturbation stream function and particle concentration, $\Psi(k_x, k_y, t)$, $\Phi(k_x, k_y, t)$, satisfy [1]

$$k^4\Psi + \alpha(k_x^2 - k_y^2)\Phi = 0 \tag{7}$$

and

$$\partial_t\Phi - k_x\partial_{k_y}\Phi = -k^2\{\nu(k_x^2 - k_y^2)\Psi + \kappa\Phi\} \tag{8}$$

where

$$k^2 = k_x^2 + k_y^2 \ \text{ and } \ \alpha = \left(\dot{\gamma}\frac{d}{d\phi}\log\eta\right)^{(0)} \tag{9}$$

and we employ the change in notation:

$$\dot{\gamma}^{(0)} t \to t, \ \frac{\kappa^{(0)}}{\dot{\gamma}^{(0)}} \to \kappa, \ \frac{\nu^{(0)}}{\dot{\gamma}^{(0)}} \to \nu \tag{10}$$

Ψ is readily eliminated from (7)-(8), the completely static form of which is elliptic for $\kappa > \nu\alpha$ and otherwise hyperbolic [1].

With Ψ eliminated, (7)-(8) become

$$\partial_t \Phi - k_x \partial_{k_y} \Phi = \sigma\{\mathbf{k}\}\Phi \tag{11}$$

where

$$\sigma\{\mathbf{k}\} = -k^2\{\kappa - \frac{\nu\alpha}{k^4}(k_x^2 - k_y^2)^2\} \tag{12}$$

As discussed in the earlier work [1], the term $k_x \partial_{k_y}$ represents "wave-vector shearing" and becomes becomes dominant for large time. In an arbitrary isochoric homogeneous flow, with $\nabla \mathbf{v} = \mathbf{L}^T$ =const. and $\mathrm{tr}\{\mathbf{L}\} \equiv \nabla \cdot \mathbf{v} = 0$, this generally can be concisely represented by the transformation:

$$\hat{\mathbf{k}} = \mathbf{F}^T \mathbf{k}, \quad \text{with } (\partial_t)_{\hat{\mathbf{k}}} = (\partial_t)_{\mathbf{k}} - (\mathbf{L}^T \mathbf{k}) \cdot \partial_{\mathbf{k}} \tag{13}$$

where

$$\mathbf{F}(t) = e^{\mathbf{L}t} \tag{14}$$

is the relevant deformation gradient.

For the simple shear at hand, (13) gives $\mathbf{k}(\hat{\mathbf{k}}, t)$ as

$$k_x \equiv \hat{k}_x \quad \text{and } k_y = \hat{k}_y - \hat{k}_x t \tag{15}$$

and converts (11) into the o.d.e.

$$\frac{d}{dt} \log \Phi = \sigma\{\mathbf{k}(\hat{\mathbf{k}}, t)\} \tag{16}$$

where $\sigma\{\mathbf{k}\}$ is given by (12).

Asymptotic stability is determined by the limiting form of σ for large t, and one finds [1] from the above that

$$\log \Phi \sim (\nu\alpha - \kappa)\frac{\hat{k}_x^2 t^3}{3} \sim (\nu\alpha - \kappa)\frac{k_y^2 t}{3}, \quad \text{for } \hat{k}_x \neq 0 \tag{17}$$

and

$$\log \Phi \equiv (\nu\alpha - \kappa)k_y^2 t, \quad \text{for } \hat{k}_x = 0 \tag{18}$$

for $t \to \infty$. Thus, $\kappa < \nu\alpha$ implies short-wavelength instability in the form of a "shear-band" structure, with $k_y >> k_x$, representing particle-depleted strata lying parallel to x.

3. Shear-Thinning Effects - Conclusions

Consider the slightly more complex ("generalized Newtonian") model

$$\mathbf{T} = 2\eta(\phi, \dot{\gamma})\mathbf{D} - p\mathbf{1} \tag{19}$$

with the same particle flux as in (2). The same type of analysis as above readily gives the modified stability criterion:

$$\kappa > \frac{\nu\alpha}{1+\beta} \tag{20}$$

where α is defined by (9), and

$$\beta = \left(\dot{\gamma}\frac{\partial \log \eta}{\partial \dot{\gamma}}\right)^{(0)} \tag{21}$$

is a local "power-law" exponent. Shear-thinning, $\beta < 0$, obviously destabilizes, with $\beta \to -1$ giving rise to to purely mechanical instability. Stability again coincides with static ellipticity.

As further work, it would be interesting to examine the possible formation of a core-annular structure in pipe flow in the unstable case. Apart from the relevance to technological applications such as lubricated pipeline transport, *e.g.* in oil-water emulsions, there may be a connection to the *Fahraeus* and *Fahraeus-Lindqvist* effects [6] associated with blood flow in small tubes.

Acknowledgements

Partial support from the U.S. National Aeronautics and Space Administration (Grant NAG 3-1888), the U.S. National Science Foundation (Grant CTS-9510121) and the 1996 Programme on the Dynamics of Complex Fluids in the Isaac Newton Institute, Cambridge University, U.K., is gratefully acknowledged.

References

1. Goddard, J.D. (1997) Migrational Instabilities in Particle Suspensions, in *The Dynamics of Complex Fluids*, Proceedings Royal Society-Unilever Indo/UK Forum, M. Adams *et al.*, eds., Imperial College Press, London (to appear).
2. Leighton, D. T. & Acrivos, A. (1987) The shear-induced migration of particles in concentrated suspensions, *J. Fluid Mech.*, **181**, 415.
3. Phillips, R.J. *et al.* (1991) A constitutive equation for concentrated suspensions that accounts for shear-induced particle migration, *Phys. Fluids A*, **4**, 30.
4. Phan-Thien, N. & Fang, Z. (1996) Entrance length and pulsatile flows of a model concentrated suspension, *J. Rheology*, **40**, 521.
5. Leal, L.G.(1980) Particle Motions in a Viscous Fluid. *Ann. Rev. Fluid Mech.*, **12**, 435.
6. Chien, S., Usami, S. & Skalak, R. (1984) Blood Flow in Small Tubes. in *Handbook of Physiology - The Cardiovascular System* , Section 2, Vol. IV, Part 1, p. 217, Renkin, E.M. & Michel, C.C., eds., American Physiological Society, Bethesda.

ON THE COMBINED EFFECT OF DIFFUSION AND AGGLOMERATION IN SHEAR FLOW OF SUSPENSIONS

H. BUGGISCH *and* G. BARTHELMES
Inst. f. Mechanische Verfahrenstechnik und Mechanik
Universität Karlsruhe
D-76128 Karlsruhe
Germany

Synopsis

In this paper, a simple model will be proposed concerning the combined effect of diffusion, agglomeration and agglomerate break-up on the rheology of shear flow. Basic underlying assumptions are:

- In shear flow, suspended particles form agglomerates and agglomerates break up. These processes are reversible.
- In homogeneous steady shear flow, the equilibrium size distribution of agglomerates is determined by the shear stress.
- In non-equilibrium situations, the rate of change of the size distribution is proportional to the distance of the state from equilibrium.
- Due to diffusion, agglomerates migrate perpendicular to the direction of flow.
- At walls, agglomerates may be built up or destroyed.
- The viscosity of the suspension depends on the agglomerate size distribution.

Possible effects of these mechanisms on the rheology of shear flow will be discussed, and it will be shown that the "viscosity" may depend on geometry. Furthermore, a simple theory of wall slip will be proposed, assuming that agglomerates are destroyed at the wall. This results in a thin wall layer with reduced viscosity, of which the thickness is determined by diffusion and the time constant of agglomeration.

1. Introduction

Rheological properties of dispersions depend on the flow-induced microstructure of these systems. In steady homogeneous shear flow, an equilibrium structure builds

up which determines the steady state viscosity of the dispersion (e.g. suspension or emulsion) [1].
If the shear flow is not homogeneous, i.e. if the shear rate is not constant throughout the fluid, structure elements (agglomerates) may migrate perpendicular to the direction of flow due to hydrodynamic diffusion or Brownian diffusion. Hydrodynamic diffusion refers to fluctuating motions of particles due to particle-particle interactions. These result in a particle flux down concentration gradients [2].
The aim of this paper is to demonstrate that diffusion, combined with flow induced local changes in microstructure, may result in interesting effects on the rheological behaviour of dispersions. For this purpose, a simple model system is considered in the following.

2. A simple mathematical model

A suspension is regarded in which particles are assumed to be either primary particles or agglomerates. Let the volume concentration of the agglomerates be q, and let the viscosity of the system depend solely on q:

$$\eta = \eta(q). \tag{1}$$

The volume concentration q is assumed to depend on the stress history. In homogeneous shear flow (i.e. if diffusion effects are absent), let the equilibrium value of q be a function of the shear stress τ only,

$$q = \tilde{q}(\tau). \tag{2}$$

We further assume that the time evolution of q without diffusion effects would be

$$\frac{\partial q}{\partial t} = -\frac{q - \tilde{q}(\tau)}{\lambda} \tag{3}$$

with a relaxation time λ.
Now consider a plane Couette flow with a shear stress τ_1 for $t < 0$ and a shear stress $\tau_2 \neq \tau_1$ for $t \geq 0$.

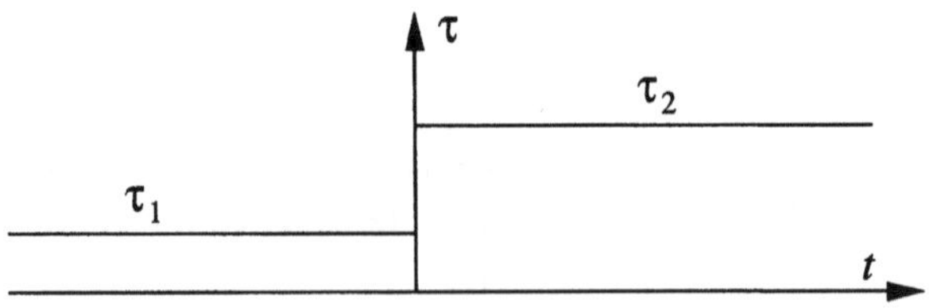

Figure 1: Step in shear stress

For $t < 0$, the structure of the suspension is the equilibrium structure for τ_1 throughout the gap:

$$q_1 = \tilde{q}(\tau_1) \tag{4}$$

The shear rate is the equilibrium shear rate:

$$\kappa = \tilde{\kappa}(\tau_1) = \kappa(\tau_1, \tilde{q}(\tau_1)) \tag{5}$$

Immediately after the step in shear stress (at $t = 0$), the structure of the suspension is still the equilibrium structure for τ_1, but the shear rate has changed [3]:

$$\kappa = \kappa(\tau_2, q_1) \tag{6}$$

At long times after the shear step ($t \gg \lambda$), a new equilibrium state is obtained with

$$q_2 = \tilde{q}(\tau_2) \tag{7}$$

and

$$\kappa = \tilde{\kappa}(\tau_2) = \kappa(\tau_2, \tilde{q}(\tau_2)) \tag{8}$$

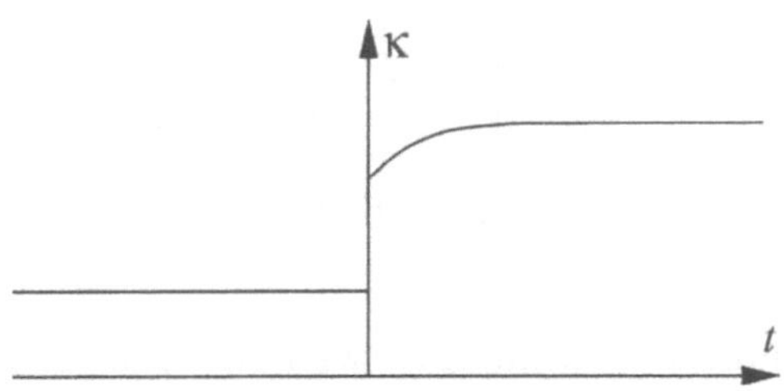

Figure 2: Shear rate reaction after step in shear stress

In non-homogeneous shear flow, e.g. in plane shear flow with finite pressure gradient (Fig.3), q will depend on y, hence diffusion of agglomerates will occur.

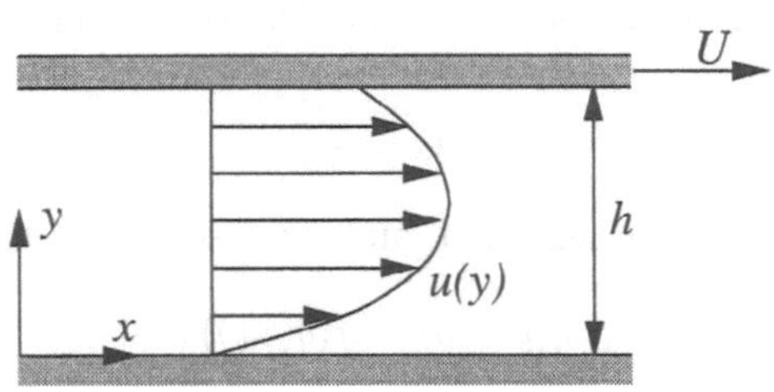

Figure 3: Plane Shear Flow

Let the agglomerate flux due to diffusive transport be $-D\frac{\partial q}{\partial y}$. In case of hydrodynamic diffusion, D is proportional to the shear rate $\kappa = \frac{\partial u}{\partial y}$, and therefore depends

on y. Additionally, agglomerates can migrate due to shear rate and viscosity gradients. For simplicity, we take D to be constant – as it would be for Brownian diffusion – and we neglect agglomerate migration leading to demixing.
The balance equation for q under these assumptions is:

$$\frac{\partial q}{\partial t} = -\frac{q - \tilde{q}}{\lambda} + D\frac{\partial^2 q}{\partial y^2}. \tag{9}$$

For the special case of plane steady shear flow, the shear stress is a linear function of y:

$$\tau(y) = \tau_w + \Lambda y. \tag{10}$$

Then the above balance equation reduces to

$$\frac{\partial^2 q(y)}{\partial y} = \frac{q - \tilde{q}(\tau(y))}{l^2}. \tag{11}$$

In this equation, $l = \sqrt{D\lambda}$ is a diffusion length scale. For the solution of this differential equation we assume the following boundary conditions:

$$D\frac{dq}{dy} = \begin{cases} +kq & at \quad y = 0 \\ -kq & at \quad y = h \end{cases} \tag{12}$$

with kq being the disagglomeration rate at the walls.

3. Results for plane Poiseuille flow without wall effects

As a first example, consider the special case $k = 0$ (no wall effect) and velocity of the upper wall $U = 0$ (Poiseuille flow). The stress distribution in this case is

$$\tau(y) = \tau_w(1 - 2\frac{y}{h}). \tag{13}$$

The properties of this flow depend on the Damköhler number (nondimensional diffusion coefficent).

$$Da = \frac{D\lambda}{h^2} \tag{14}$$

If $Da \ll 1$, diffusion can be neglected, and the viscosity at the wall is

$$\eta(y = 0) = \eta\left(\tilde{q}(\tau_w)\right). \tag{15}$$

This is the viscosity measured by a slit rheometer for $Da \ll 1$. The same viscosity would be obtained in a plane Couette rheometer with $\tau = \tau_w$, independent of Da. If on the other hand $Da \gg 1$, diffusion would result in a homogeneous agglomerate distribution with

$$q = \bar{q} = \frac{1}{h}\int_0^h \tilde{q}(\tau(y))dy. \tag{16}$$

In this case, the viscosity measured by a slit rheometer would be

$$\eta = \eta(\bar{q}). \tag{17}$$

In general, the value obtained for the viscosity at the wall will depend on Da and therefore on geometry:

$$\eta(y=0) = \eta(Da, \tau_w). \tag{18}$$

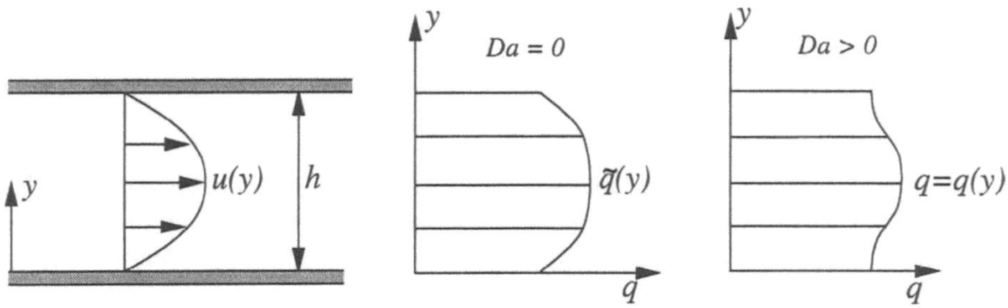

Figure 4: Influence of diffusion on the concentration profile in plane Poiseuille flow

4. Pseudo wall slip in plane Couette flow

Now assume $\Lambda = 0$ and $\tau = \tau_w = const$ (Couette flow). As long as $k = 0$ (no disagglomeration at the wall), the flow is homogeneous with constant agglomerate concentration throughout the gap. In equilibrium, the viscosity is $\eta(\tilde{q}(\tau_w))$ everywhere. Then consider the case $k > 0$, in which agglomerates are destroyed at the wall. Far away from walls, q will have its equilibrium value $\tilde{q}(\tau_w)$. Close to the wall, a boundary layer will develop due to the combined effects of wall disagglomeration, diffusion and reagglomeration in the bulk of the suspension. For this flow, the balance equation (11) can be solved using the boundary condition (12):

$$q = \tilde{q}(\tau_w)\left(1 - \frac{l}{l + \frac{D}{k}} e^{-y/l}\right). \tag{19}$$

The thickness of the wall layer is of the order $O(l)$. Immediately at the wall, q assumes the value

$$q(y=0) = \tilde{q}(\tau_w)\frac{1}{1 + \frac{k}{D} l}. \tag{20}$$

Due to wall disagglomeration, the viscosity $\eta(q(y))$ within the boundary layer is lower than the equilibrium viscosity $\eta(\tilde{q})$, which results in a "pseudo wall slip" of the dispersion.

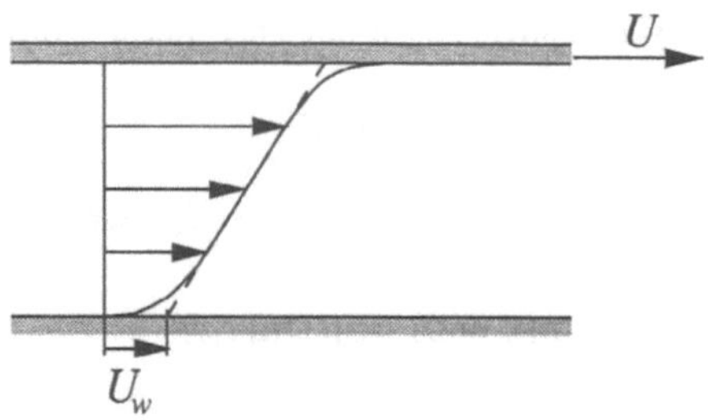

Figure 5: Pseudo wall slip

5. Conclusional remarks

A simple mathematical model has been proposed showing the principal ideas for the combined effect of diffusion, agglomeration and agglomerate break-up on the rheology of suspensions. Many simplifications have been made, which neglect known relations and phenomena.

In suspension flow with hydrodynamic diffusion, the diffusion coefficient is depending on shear rate κ and particle radius a [4]: $D \sim a^2|\kappa|$.

In addition to the diffusive flux due to concentration gradients, there is a particle flux due to gradients in shear rate and viscosity, which balances the flux due to concentration gradients in steady state [5].

The disagglomeration rate at the walls may be a function of the wall shear stress, $k = k(\tau_w)$, and the agglomeration kinetics may depend on the local shear rate, $\lambda = \lambda(\kappa)$. In subsequent work, these influences have to be considered.

References

[1] Reiner, M.: Advanced Rheology, H.K. Lewis, London (1971)

[2] Davis, R.H.: Hydrodynamic diffusion of suspended particles: a symposium, J. Fluid Mech. *310* (1996) pp 325-335

[3] Windhab, E.: Untersuchungen zum rheologischen Verhalten konzentrierter Suspensionen, Dissertation, Universität Karlsruhe (1986)

[4] D. Leighton, A. Acrivos: The Shear-induced Migration of Particles in Concentrated Suspensions, J. Fluid Mech. *181* (1987) pp 415-439

[5] R.J. Phillips, R.C. Armstrong, R.A. Brown, A.L. Graham, J.R. Abbott: A Constitutive Equation for Concentrated Suspensions that Accounts for Shear-induced Particle Migration, Phys. Fluids A *4* (1992) pp 30-40

STEADY FLOW OF A POWER LAW FLUID THROUGH AN ELASTICALLY DEFORMABLE POROUS MEDIA

B. S. BHATT

Department of Mathematics and Computer Science
The University of the West Indies,
St. Augustine, Trinidad (W.I.)

The steady one dimensional flow of a powerlaw fluid through a deformable porous media has been studied. The governing equations have been solved for various constitutive relations relating stiffness and permeability of the porous matrix.

1. Introduction

In the present study we extend the analysis of Parker et. al. (1987) for a powerlaw fluid. One dimensional flow of a powerlaw fluid has been considered through a porous media which is deformable.

Caro et. al. (1984) have studied the transport of fluid through deformable porous materials with relevance to connective tissue properties. In general the nature of the flow and the resultant strain distribution depend upon the properties of the porous media and the fluid and their interaction as well as the boundary conditions geometry and applied pressure etc.

The Theories of flow in porous, deformable media were first developed to explain the consolidation phenomena in soil mechanics [see Terzaghi (1925], later on the equations were proposed by Biot (1955) which are widely accepted and are used to explain observed behaviour in geological materials. These equations have been supported by Thermodynamic Theory of mixtures by Crochet and Naghdi (1966), Rice and Cleary (1976), Kenyon (1976 a, 1976 b) and Burridge and Keller (1981).

More recently the interest in flow through deformable porous media has been stimulated by the problems in Biomechanics [see Lai and Mow (1980), Lai et. al. (1981), Kenyon (1979) and Jayaraman (1983)]. Some of the areas of interest can be in land sliding, ground water flows, human and animal joints where we come across flow of fluids through porous deformable media.

Recently Jared and Andreas (1996) have discussed the infiltration of an initially dry deformable porous media by a pressurized liquid taking the influence of variation in permeability into account.

2. Mathematical Model

(a) *Basic Equation*

We consider one dimensional motion of a powerlaw fluid through a deformable porous material. Basically the equations are of Biot (1955) with definitions and interpretation of Kenyon (1976 a, 1976 b). In this continuum approach the solid and fluid co-exist. Introducing non-dimensional quantities

$$w = \frac{u}{\rho m / Ho}, \qquad x = \frac{z}{Ho}, \quad p = \frac{P}{\rho^3 (m / Ho)^2}$$

$$\lambda = \frac{\Lambda}{\rho^3 (m / Ho)^2}, \quad \tau = \frac{t}{Ho^2 / \rho m}, \quad \phi = \frac{k}{Ho^2 \alpha}\left(\frac{\rho m}{Ho}\right)^{n-1},$$

the momentum equation for the solid matrix and the Darcy's law give

$$-\frac{1}{\phi}w^n = \frac{\partial p}{\partial x} = \frac{\partial(\lambda R)}{\partial x}\ , \tag{1}$$

$$\frac{\partial w}{\partial x} = -\frac{\partial R}{\partial \tau}\ , \tag{2}$$

where w, p, λ, R, ϕ and τ are respectively dimensionless velocity, pressure, stiffness, local strain gradient, permeability and time, $n > 1$ for dilatant fluid, $n = 1$ for Newtonian fluid and $n < 1$ for pseudoplastic fluid.

For steady flow, equations (1) and (2) give

$$\frac{d}{dx}\left(\phi\frac{d}{dx}(\lambda R)\right) = 0\ . \tag{3}$$

The solid is described by constitutive equations which relate the stiffness, permeability and the local strain gradient.

(b) *Boundary conditions*

We consider a homogeneous matrix of undeformed thickness Ho restrained at $x = 0$ by a rigid grid which does not impede the flow. A constant pressure p_1 is maintained at the bottom and a surface stress s_1 is applied both at top and bottom of the porous matrix, therefore we have

$$\left.\begin{array}{l} x = 0,\ \ \lambda R = s_1 + p_1, \\ x = h,\ \ \lambda R = s_1 . \end{array}\right\} \tag{4}$$

The matching condition is taken to be

$$\int_0^h R\,dx = 1 - h, \tag{5}$$

for $y = x/h$, (5) gives

$$h = \left[1 + \int_0^1 R\,dy\right]^{-1}\ . \tag{6}$$

(c) *Steady state*

If the applied stresses p_1 and s_1 are constants, then the problem is steady and the equations (1) - (4) and (6) are sufficient to determine w and h.

3. RESULTS

In general solutions are possible for K(R) and $\lambda(R)$.
We look at solutions for various relations of these functions.

(a) *Zero flow*

If $p_1 = 0$, from equations (1) and (4) $R = s_1 / \lambda$ substituting in equations (6), we get

$$h = \left[(s_1 / \lambda) + 1\right]^{-1} \tag{7}$$

For rest of the analysis we take $s_1 = 0$.

(b) *For constant permeability*

We take $\phi = 1$ and consider

(i) *Constant Stiffness*

We take $\lambda = 1$. The equation (3) gives

$$R = p_1(1-y), \tag{8}$$

where $y = x / h$.

Therefore equation (6) gives

$$h = (1 + p_1 / 2)^{-1}, \tag{9}$$

and we write w as w_1 which is given by

$$w_1 = \left[p_1(1 + p_1 / 2)\right]^{\frac{1}{n}} \tag{10}$$

The behaviour of w_1 versus p_1 has been given in figure 1 for $n = 0.8$ (pseudoplastic fluid), $n = 1$ (Newtonian fluid) and $n = 1.2$ (dilatant fluid). We see that w_1 increases with p_1 and as n decreases.

(ii) *Exponential stiffness*

We take $\lambda = a + (1-a)\exp(-bR)$, (11)
where a is the limiting stiffness at large compaction. From the boundary conditions, the strain varies from o at the top at the matrix to R_0 at the bottom. Therefore $R_0\,\lambda(R_0) = p_1$. For small values of a $\left[a < (e^2+1)^{-1}\right]$ care must be taken because there is a range of pressures for which we get more than one value of R_0. Such a behaviour could give rise a sudden collapse of the porous matrix.

In this case we obtain

$$\int_0^1 R\,dy = \frac{1}{p_1}\left[\frac{1}{b^2}(p_1/R_0+1) - \frac{R_0}{b}(a - p_1/R_0) + R_0^{\,2}(p_1/R_0 - a/2)\right], \tag{12}$$

Therefore the velocity (we write $w = w_2$) is obtained as

$$w_2 = (p_1/h)^{\frac{1}{n}}, \tag{13}$$

where h is given by the equation (6). Figures 2-4 give the behaviour of w_2 with p_1 for various values of a when n =0.8. For small values of a = 0.01 and a = 0.1 we get hysteresis due to the multivalued nature of the assumed stress-strain law. The limiting case, a =1, is the solution for constant stiffness and permeability. Figures 3 and 4 give magnified view of figure 1 at a =0.01.

(c) *Constant Stiffness*

We take $\lambda = 1$. Then the equations (3) and (4) reduce to

$$\phi(R)\frac{dR}{dx} = -w^{n}, \tag{14}$$

and

$$\left.\begin{aligned} x = 0,\ R = p_1, \\ x = h,\ R = 0. \end{aligned}\right\} \tag{15}$$

We consider the following cases:

(i) *Constant permeability*

when $\phi = 1$. The results are already obtained in (i) of (b).

(ii) *Linear permeability*

We assume $\phi(R) = 1 - R/\hat{R}$, (16)

for $R < \hat{R}$, The critical strain at which the permeability function ϕ becomes zero.

In this case equations (15) - (17) give (we unite $w = w_3$),

$$R = \hat{R}\left[1-\left(1-\frac{2w_3{}^n h}{\hat{R}}(1-x/h)\right)^{\frac{1}{2}}\right], \tag{17}$$

$$w_3{}^n = p_1\left[1+\frac{p_1\left(\hat{R}-1\right)}{2\hat{R}}-\frac{p_1{}^2}{3\hat{R}}\right], \tag{18}$$

and

$$h = \frac{p_1}{w_3{}^n}\left(1-\frac{p_1}{2\hat{R}}\right). \tag{19}$$

The behaviour of w_3 with p_1 has been given in figures 5-7. The values of critical value $\hat{R}$ have been taken as R_1, R_2, R_3, R_4 & R_5. The velocity goes through maximum as pressure is increased and eventually becomes zero. This limit is probably not very realistic since real materials will deviate from the linear permeability law at large deformations. However, the flow limitation properties implied by the maximum are realistic as they occur at low levels of distortion of the porous matrix. The velocity increases with n.

(iii) *Exponential permeability*

We assume $\phi(R) = \exp(-cR)$. (20)

Then the strain R, velocity w_4 and height h are given by

$$R = -\frac{1}{c}\ln\left[1-cw^n{}_4h(1-x/h)\right], \tag{21}$$

$$w^n{}_4 = \frac{1+c}{c^2}\left[\left(1-e^{-cp_1}\right)-\frac{p_1}{c}e^{-cp_1}\right], \tag{22}$$

and

$$h = \left[\frac{c+1}{c}-\frac{p_1 e^{-cp_1}}{1-e^{-cp_1}}\right]^{-1}. \tag{23}$$

Figures 8-10 give the velocity profiles for exponential permeability law for various values of c when n = 0.8, 1.0, 1.2. The curves are monotonic but approach to asymptotic maximum.

(iv) *Inverse law of permeability*

We take $\phi(R) = \dfrac{1}{1+\alpha R}$. (24)

The velocity w_5 and height h are given as

$$w_5 = \left[\frac{p_1}{\alpha} + \frac{\alpha - 1}{\alpha^2}\ln(1+\alpha p_1)\right]^{\frac{1}{n}} \quad , \tag{25}$$

$$\text{and } h = \frac{1}{\alpha w_5^{\,n}}\ln(1+\alpha p_1). \tag{26}$$

The velocity distribution has been shown graphically in figures 11-13. The velocity increases with n.

4. References

Biot, M. A. (1955) Theory of elasticity and consolidation for a porous anisotropic solid, J. Appl. Physics.**26**, 182-185

Burridge, R. and Keller, J.B. (1981) porelasticity equations derived from microstructure, J. Acoust.Soc.Am. **70**, 1140-1146.

Caro, C.G., Lever, M.J., Morgan, G.E., Parker, K.H. and Winlove, C.P. (1984) Studies of transport in deformable porous materials with relevance to connective tissue properties, J. Physiol, **358**, 3.

Crochet, M.J. and Naghdi, P.M. (1966) On constitutive equations for flow of fluid through an elastic solid, Int. J. Engng.Soc. **4**, 383-401.

Jared, L.S. and Andreas, M. (1996) Forced unidirectional infiltration of deformable porous media, J. Fluid Mech. **311**, 193-217.

Jayaraman, G. (1983) Water transport in the arterial Wall - a theoretical study, J. Biomech. **16**, 833-840.

Kenyon, D.E. (1976a) Thermostatics of solid-fluid mixtures, Arch. Ration. Mech. Analys. **62**, 117-129.

Kenyon, D.E. (1976b) The theory of incompressible solid-fluid mixture, Arch. Ration.Mech. Analys. **62**, 131-147.

Kenyon, D.E. (1979) A mathematical model of water-flux through aortic tissue, Bull.Math.Biol. **41**, 79-90.

Lai, W.M. and Mow, V.C. (1980) Drag induced compression of articular cartilage during a permeation experiment, Biorheol.**17**, 111-123.

Lai, W.M., Mow, V.C. and Roth, V. (1981) Effects of nonlinear strain dependent permeability and rate of compression on the stress behaviour of articular cartilage, J. Biomech. Eng. **103**, 61-66.

Parker, K.H., Mehta, R.V. and Caro, C.G. (1987) Steady flow in porous, elasticity deformable materials, Trans ASME J. Appl. Mech. **54**, 794-800.

Rice, J.R. and Cleary, M.P. (1976) Some basic stress diffusion solutions for fluid saturated elastic porous media with compressible constituents. Rev. Geophys.Space phys., **14**, 227-241.

Terzaghi, K. (1925) Erdbaumechnaik auf Bodenphsikalischen Grundlagen, Wien, Deuticke.

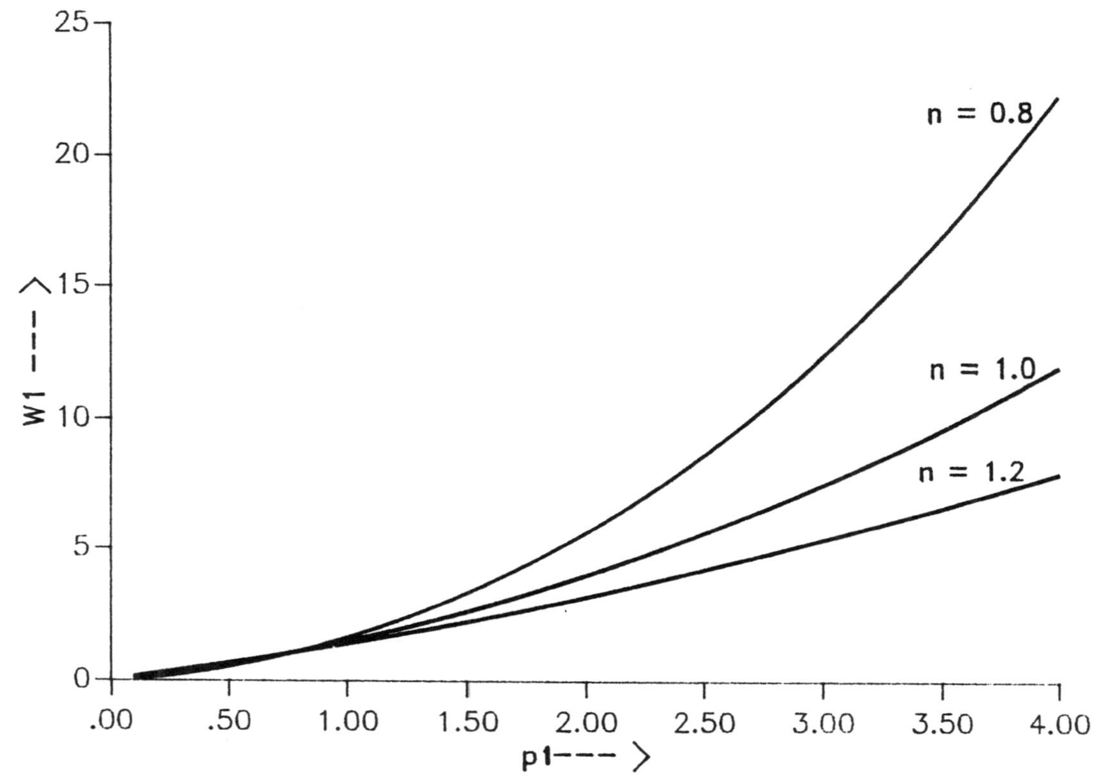

Fig. 1 w_1 versus p_1

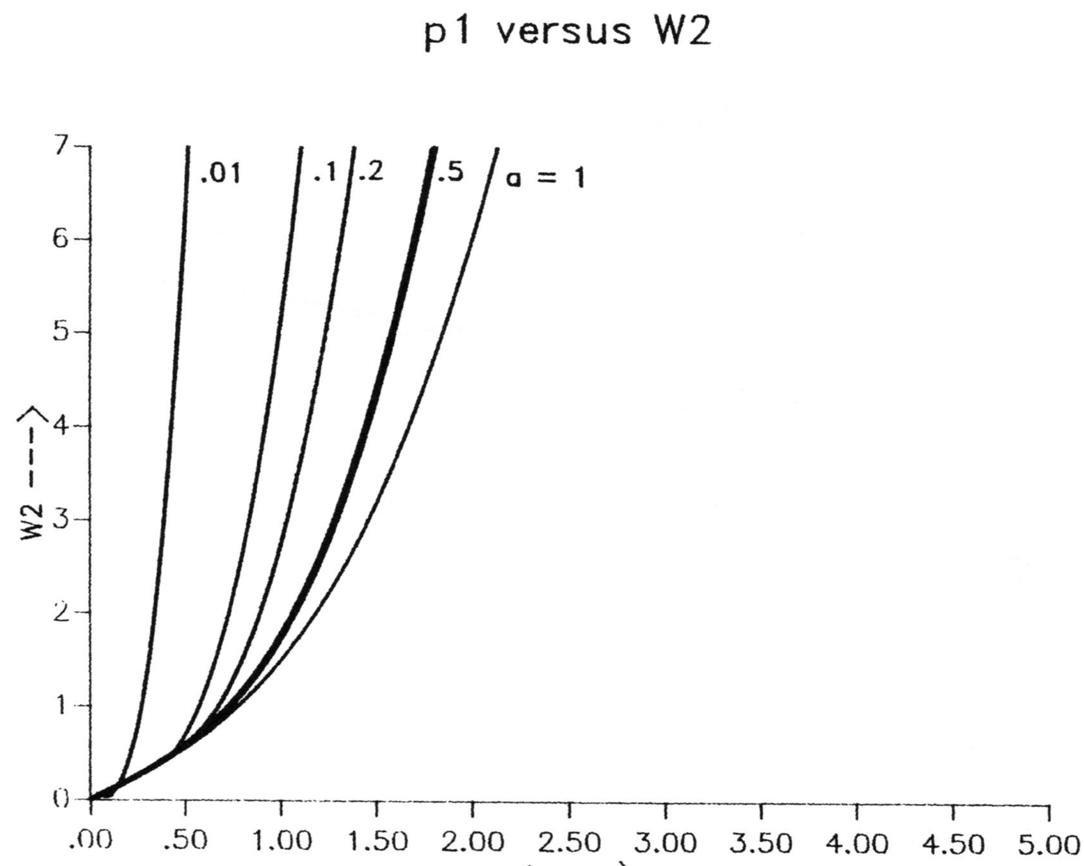

Fig.2 w_2 versus p_1

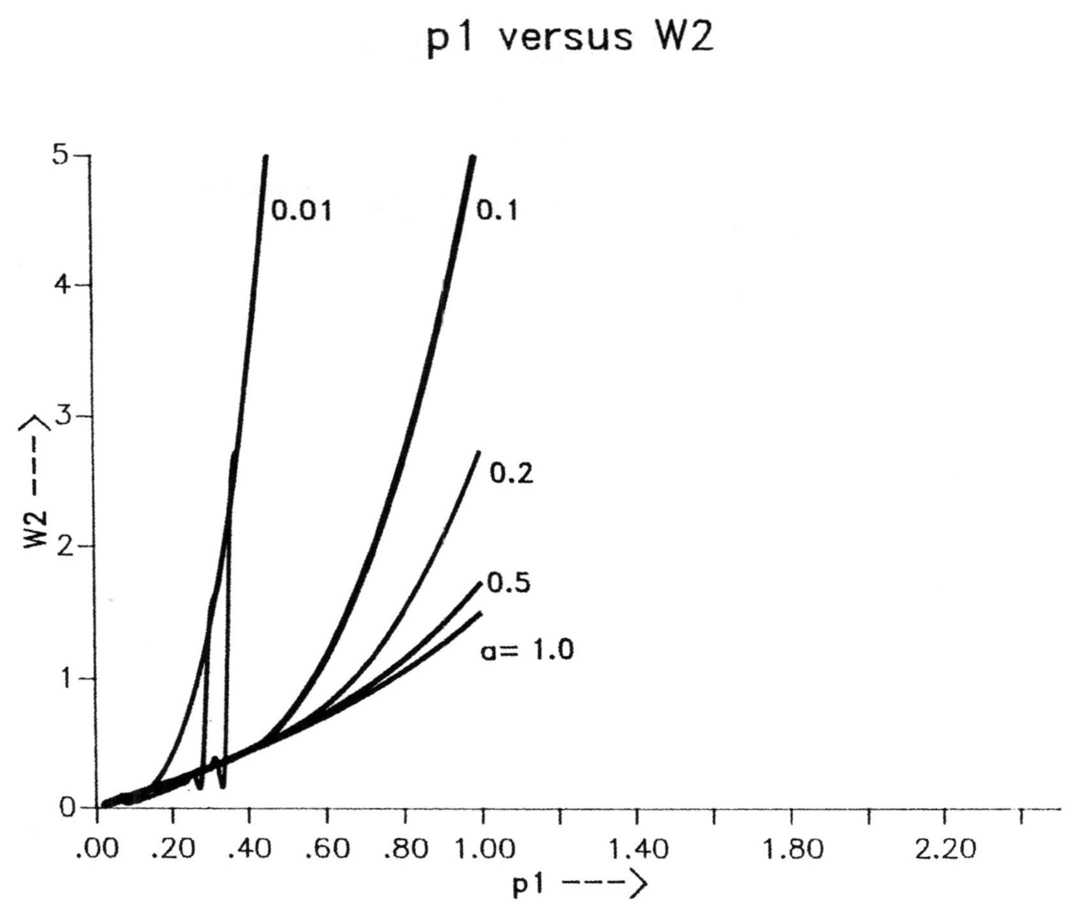

Fig.3 w_2 versus p_1

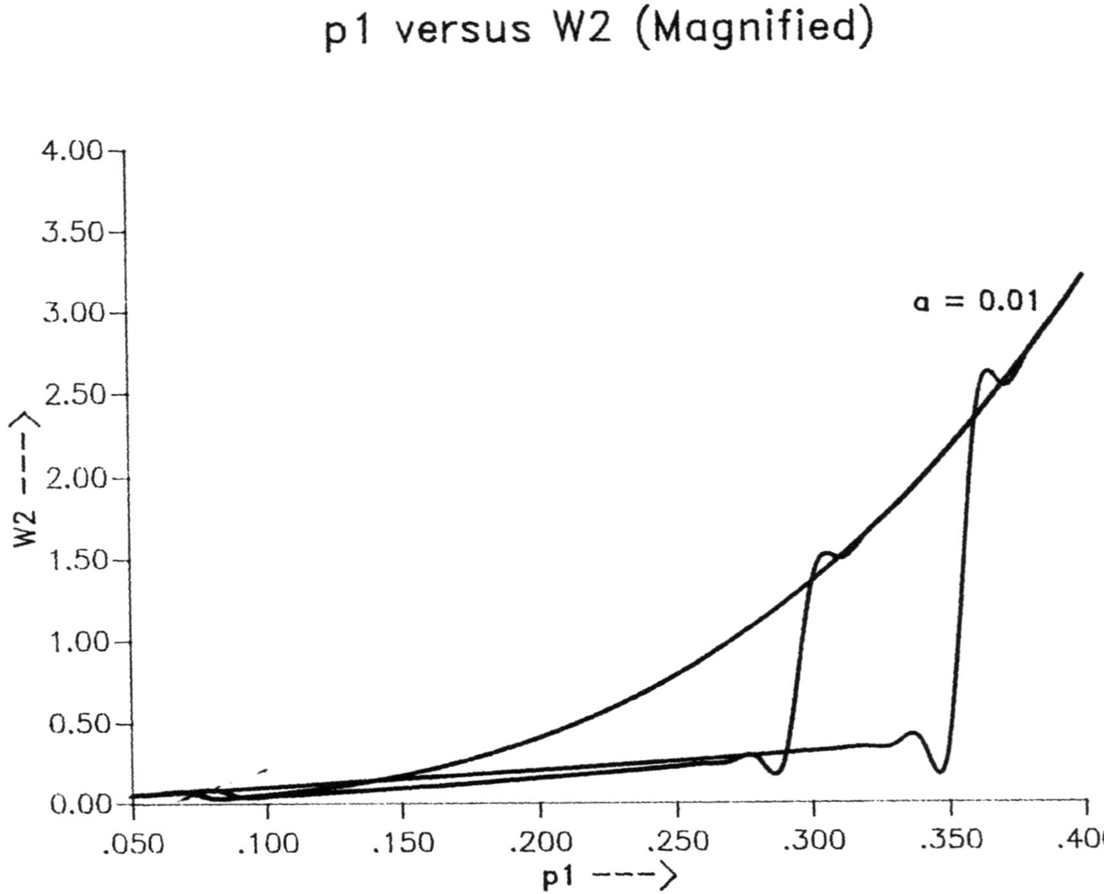

Fig.4 w_2 versus p_1

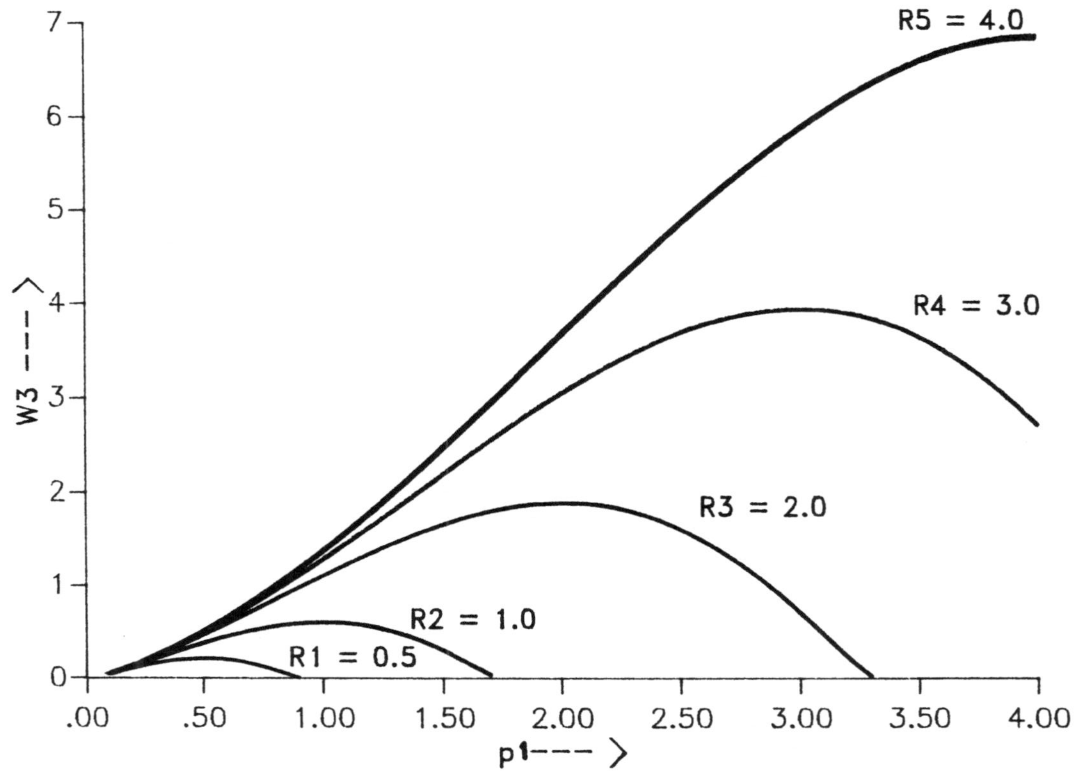

Fig.5 w_3 versus p_1 ($n = 0.8$)

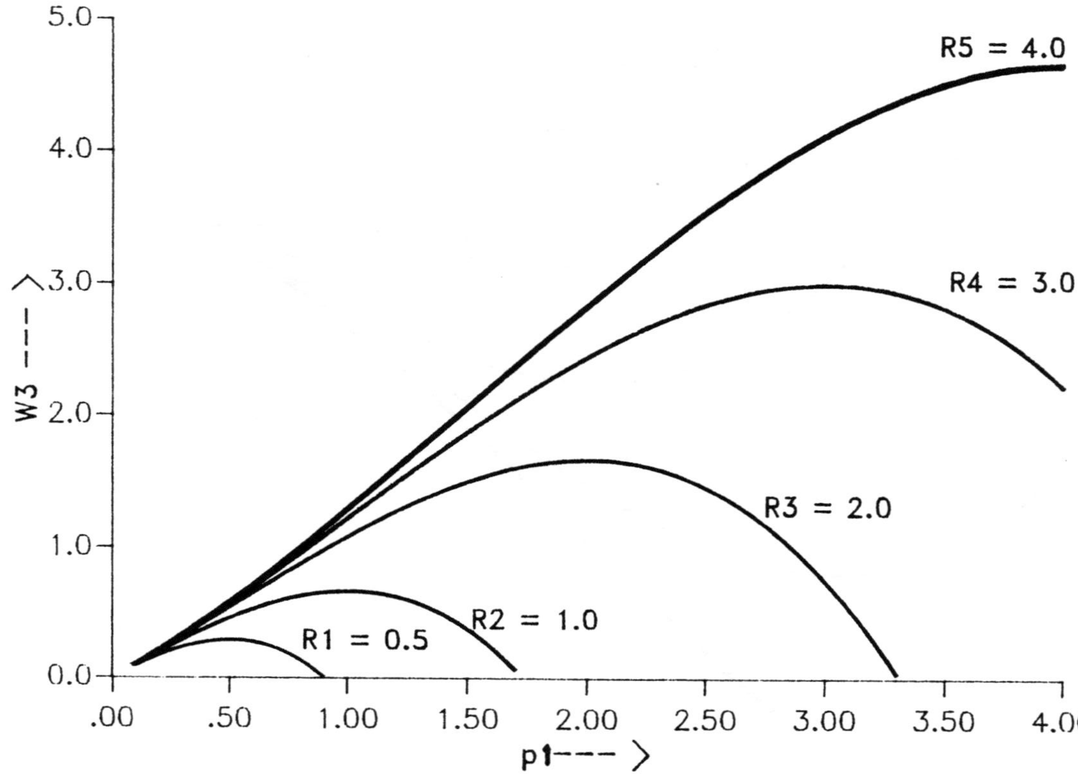

Fig.6 w_3 versus p_1 $(n = 1.0)$

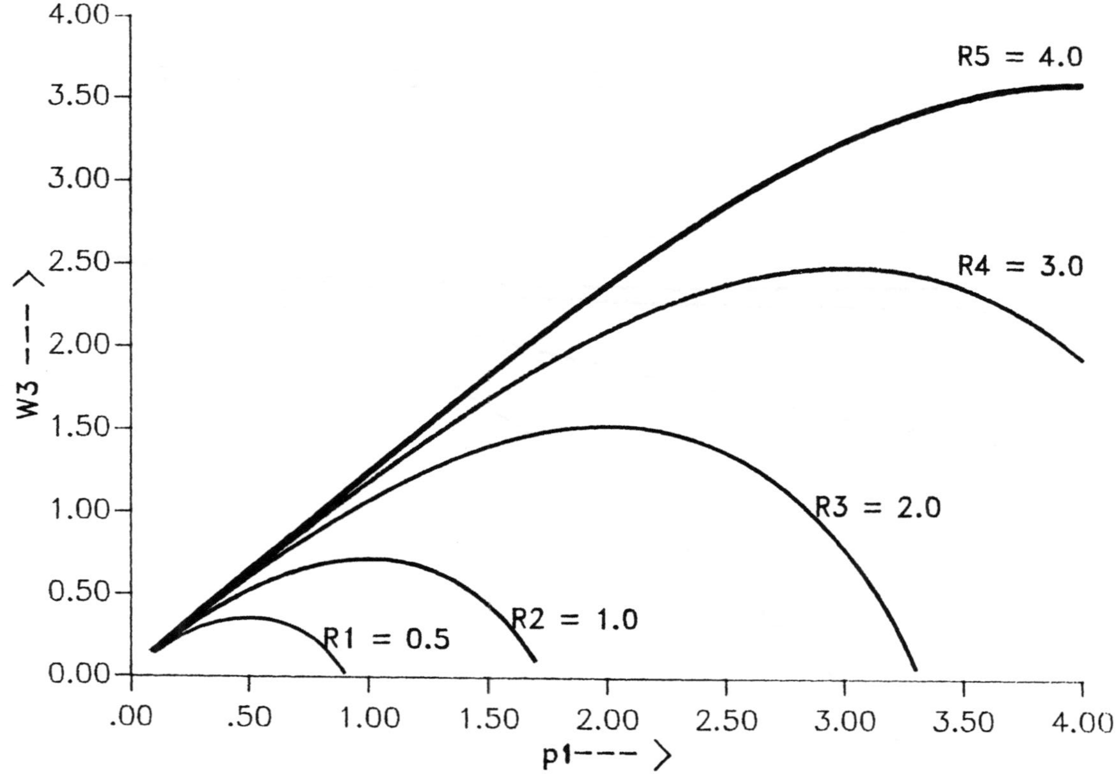

Fig.7 w_3 versus p_1 ($n = 1.2$)

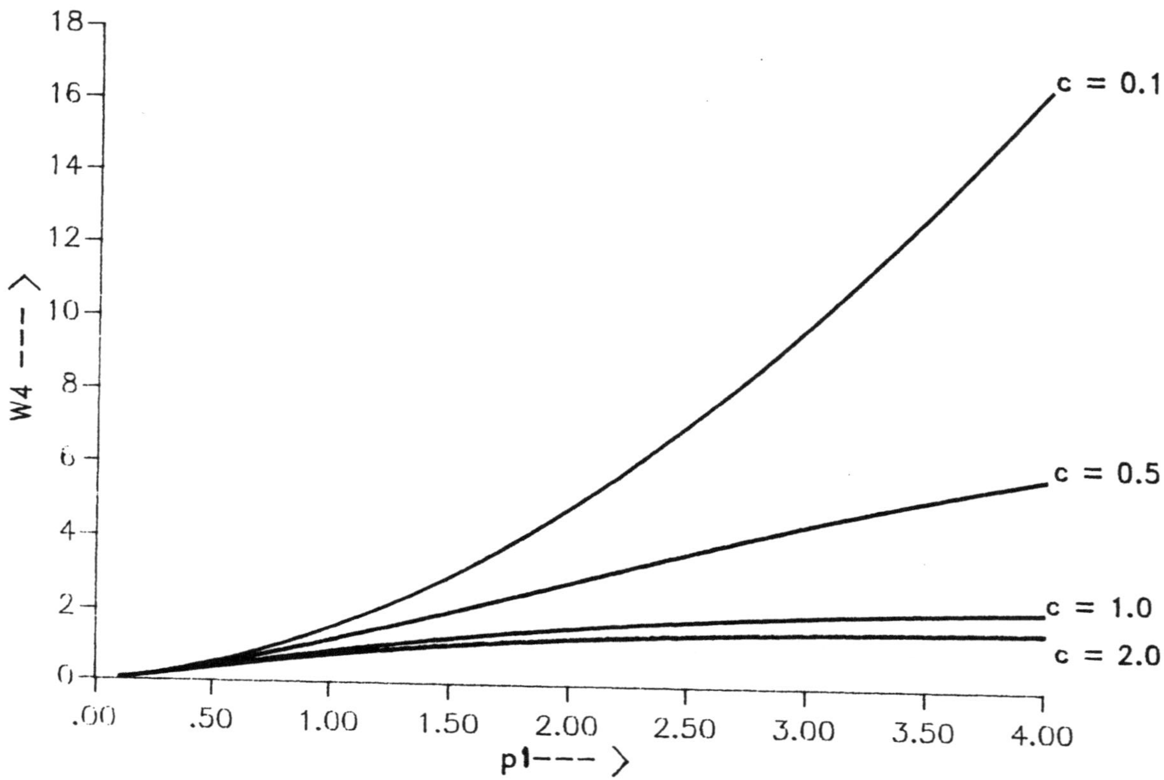

Fig.8 w_4 versus p_1 $(n = 0.8)$

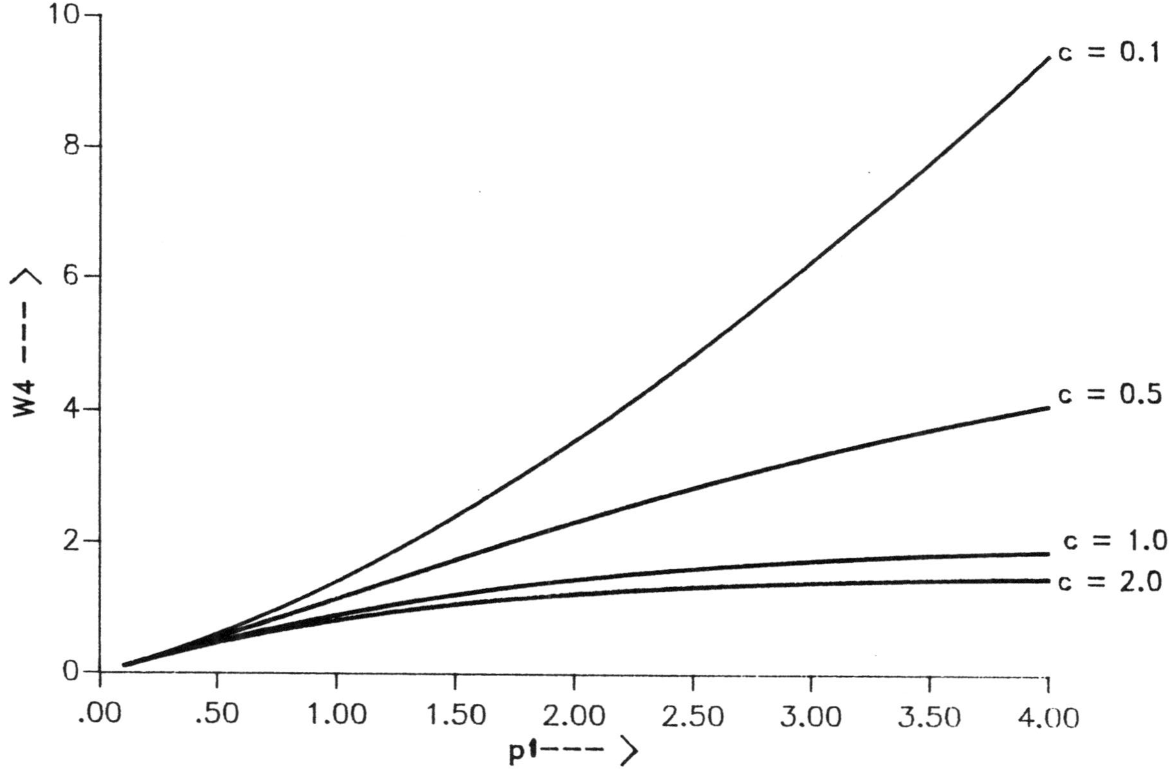

Fig.9 w_4 versus p_1 ($n = 1.0$)

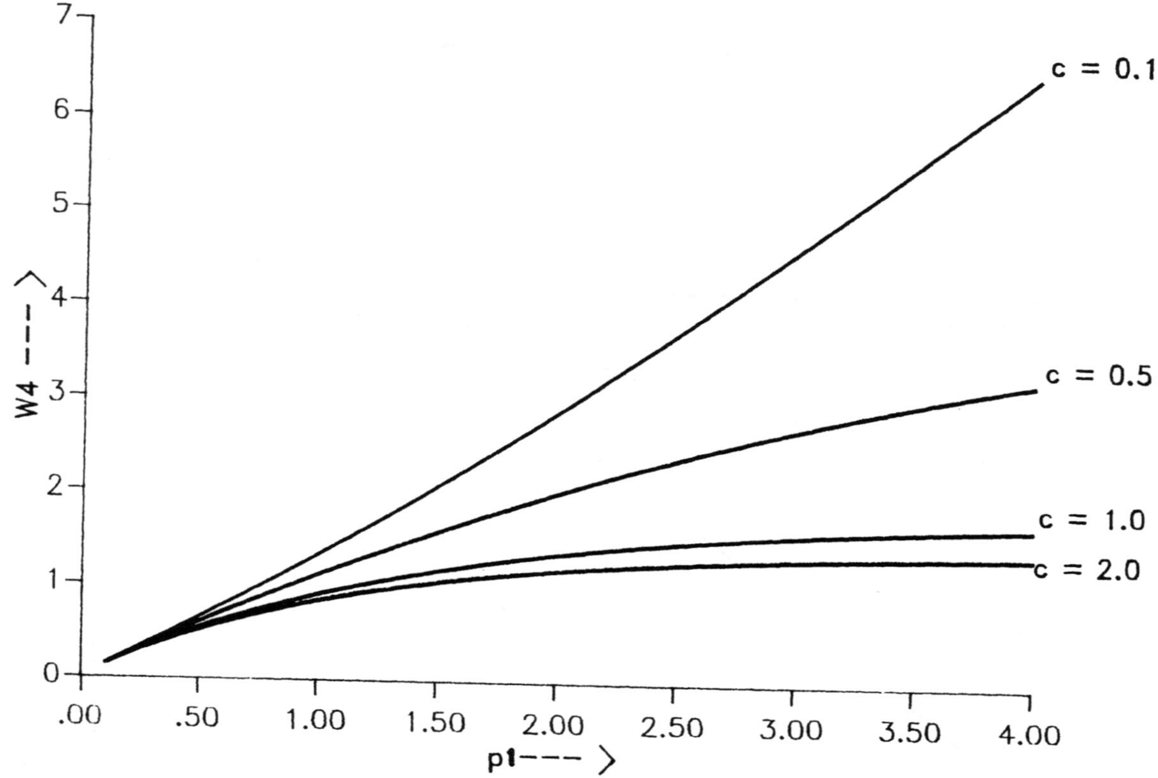

Fig.10 w_4 versus p_1 $(n = 1.2)$

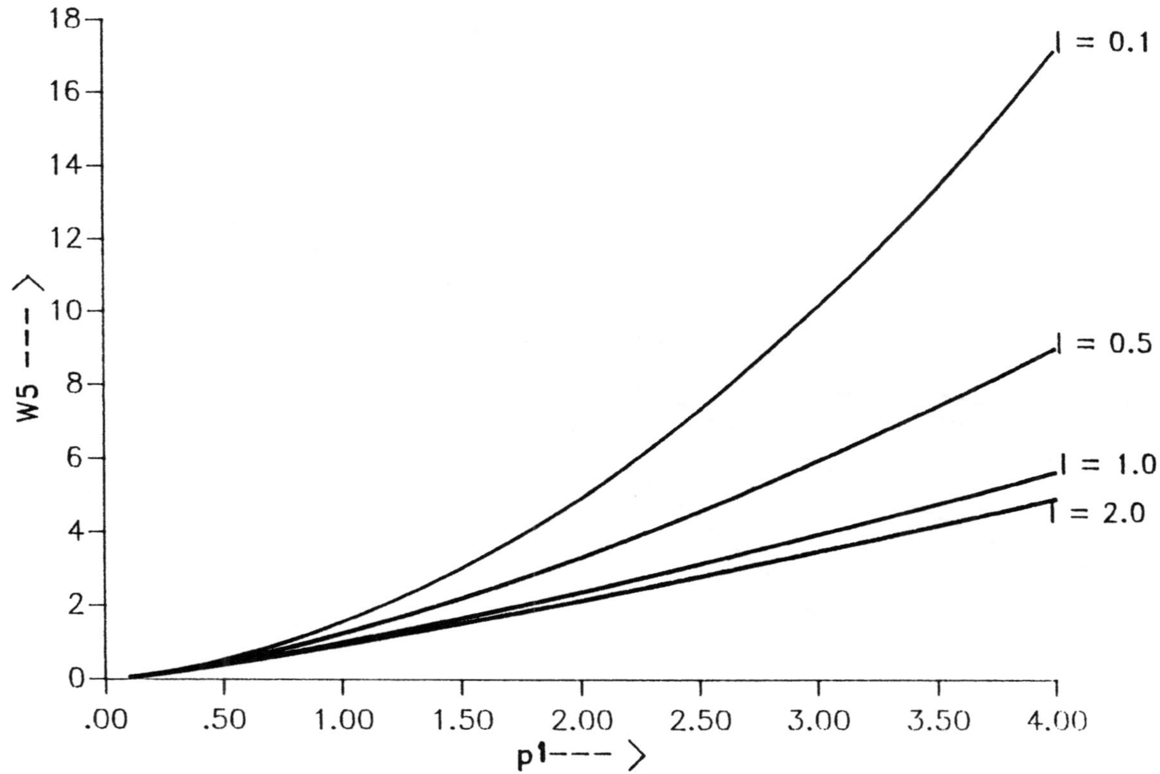

Fig.11 w_5 versus p_1 $(n = 0.8)$

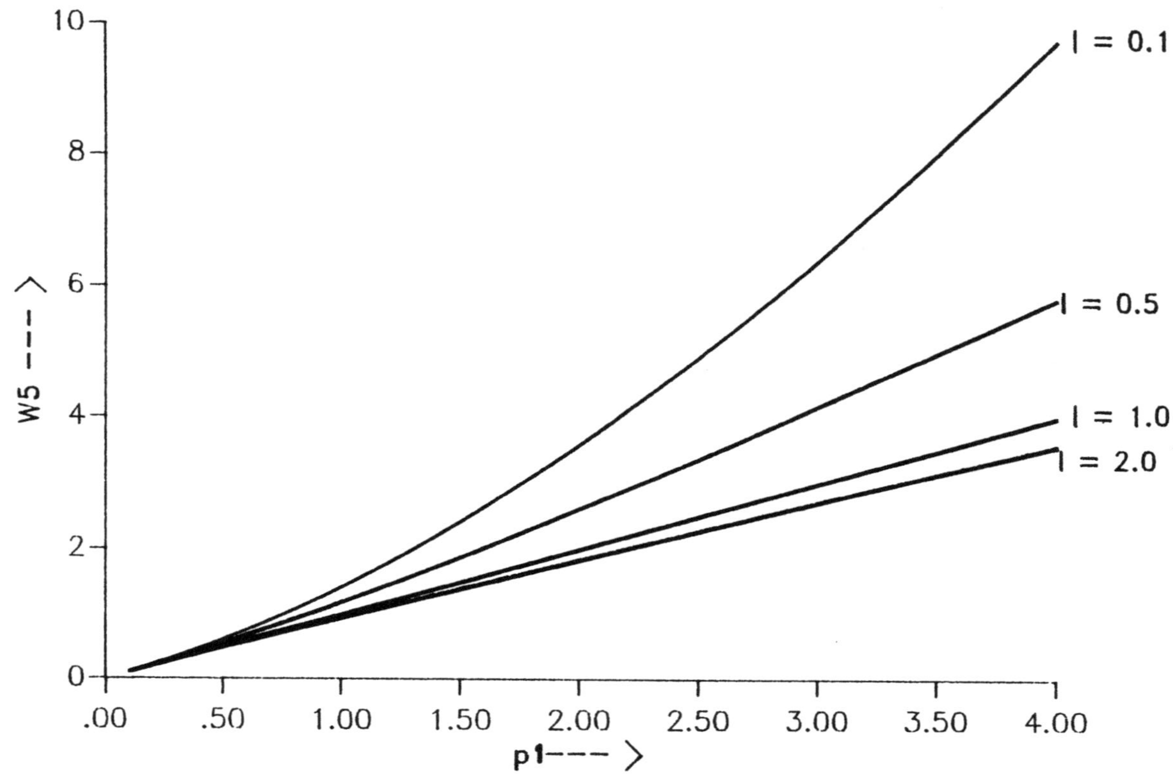

Fig 12 w_5 versus p_1 ($n = 1.0$)

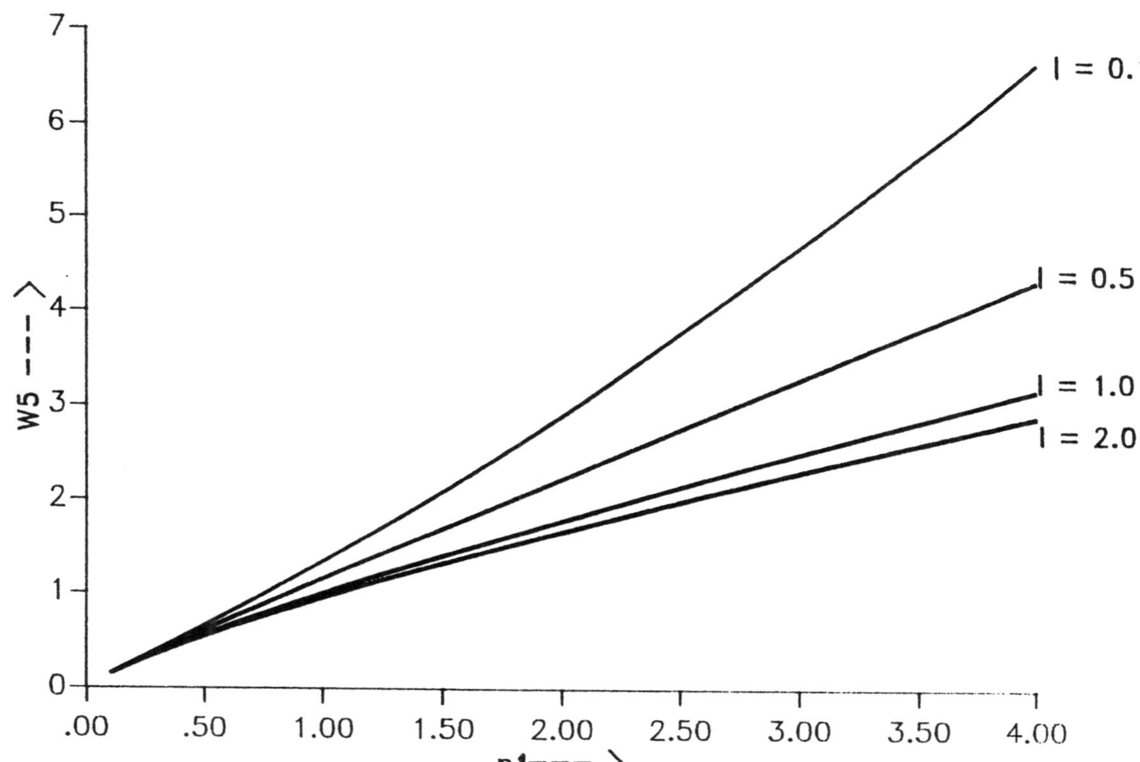

Fig.13 w_5 versus p_1 $(n = 1.2)$

THOUGHTS OLD AND NEW ON MOVING CONTACT LINES

J. R.A. Pearson & J. Ferguson
Schlumberger Cambridge Research
High Cross, Madingley Road
Cambridge CB3 0EL
England

1 Introduction

The motion of a contact line, defined as the line where the interface between two fluids meets a solid surface, over the same surface introduces a difficulty (an unacceptable singularity in deformation rate and stress fields at the contact line) into the conventional assumptions of continuum (Newtonian fluid) mechanics. This arises because traditionally it is assumed that, even at a perfectly smooth plane solid surface, there is no velocity discontinuity between fluid and solid; this is the celebrated no-slip assumption.

This assumption has been questioned by various authors. One of the most radical contributions (Richardson [9]) noted that the zero shear stress assumption of inviscid hydrodynamics, combined with small asperities on the surface would lead to an apparent no-slip condition in the case of viscous fluids. Using the example of a periodically corrugated surface Richardson showed that the apparent paradox was associated with the position of the surface on which the no-slip condition could be assumed to act. This effective surface lay within the solid bounding the fluid; if the more usual mean geometrical planar surface was used then the relevant boundary condition would be of Navier type

$$\mathbf{u} = k\partial\mathbf{u}/\partial n \tag{1}$$

with k a length determined by the geometry of the asperities and of the same order of magnitude. This simple alternative formulation is enough to remove the singularity at the contact line, [5].

Other authors have argued that the fluid in the near neighbourhood of the material surface will have different physical properties, e.g. viscosity and density, from the bulk fluid (particularly in the case of polymeric liquids) and so this too can lead to (not necessarily positive) contributions to k in equation 1 above. Shikhmurzaev [11] has extended this concept to include rate processes. Behind all these arguments lies a basic concept inherent in continuum mechanics: the molecular nature of matter is eliminated by certain averaging processes (often unspecified in detail). Moreover, short range effects in interfacial regions are subsumed into boundary conditions at interfaces that are taken to be of zero thickness: this implicitly sets a lower bound on the length scale which can be adequately modelled by continuum theories. This same argument must therefore apply to the position of the contact line itself. Thus if k in equation 1 is of the same order as molecular separations (or mean free paths in the case of gases), a continuum theory using the no-slip boundary condition is not able to consider such length scales anyway.

With reference to table 1, Pearson [8] set out (in descending orders of magnitude) the actual values of typical length scales for contact lines associated with drops of oil or water moving over obstructions (e.g. sharp probes) placed in their path, where the non-dimensional groups are defined as follows:

$$Bd = \Delta\rho g L^2/\sigma \quad We = \rho U^2 L/\sigma \quad Re = \rho U L/\mu$$

and the physical parameters have taken typical values: $\rho = 10^3$kg/m^3, $U =$ 1m/s, $\mu = 10^{-3}$ Pas, $\sigma = 10^{-2} - 10^{-3}$ Pam.

LENGTH SCALES		m
pipe width		10^{-1}
capillary length	$Bd = 1$	10^{-2}
bubble/drop size		$10^{-3} - 10^{-2}$
probe-tip size		$10^{-4} - 10^{-3}$
inertial length	$We = 1$	$10^{-5} - 10^{-4}$
asperities/chemical roughness		$10^{-9} - 10^{-5}$
viscous length	$Re = 1$	10^{-6}
wavelength of light		$10^{-7} - 10^{-6}$
surfactant molecule		$10^{-9} - 10^{-7}$
long-range forces/disjoining pressure		$10^{-9} - 10^{-7}$
short-range forces (molecular)		10^{-9}

Table 1: Length scales for contact lines associated with drops of oil moving over obstructions

Taking the capillary number

$$Ca = \mu U/\sigma \tag{2}$$

where μ is the viscosity, U is contact line speed and σ is interfacial tension, we note that $Ca = 1$ when $U = 10$ m/s for the extreme but relevant values of $\mu = 5$ x 10^{-4} Pa s and $\sigma = 5$ x 10^{-2} Pa m. This can be interpreted as indicating that the local radius of curvature of the interface will then not be small compared with distance from the contact line. This in turn means that a contact angle can be regarded as a function of (averaging) length scale. This is discussed in much more detail later.

This same velocity (10 m/s) gives a unit value of the Reynolds Number

$$Re = \rho U l/\mu \tag{3}$$

when $l = l_1 = 2\text{x}10^{-7}m$ for $\rho = 1000$ kg/m^3. From this we can deduce that inertial effects can be as important as viscous effects in the near neighbourhood of moving contact lines right down to a length scale over which long-range forces can act, or comparable with that of a stretched surfactant polymer molecule, or that of asperities in an otherwise optically flat plane.

We make this point here because almost all work on the topic of moving contact lines (Dussain [4], Cox [2], Hocking [5]) has assumed that Reynolds numbers are very small in a continuum neighbourhood of the contact line. The singular viscous solution for flow at a contact line, in any continuum theory for $Ca < 1$ (finite curvature of the interface), leads to stress gradients that behave as r^{-1}, where r is the distance from the contact line, and a pressure that therefore behaves as $\ln r$, (Huh& Scriven [6]). If the value of r is truncated at a molecular length scale, we find that the localised stress, non-dimensionalised with the wall shear stress at $l = 10^{-6}$ m, will be of order $\ln(r/l)$, i.e. $|\ln(10^{-9}/10^{-6})| \approx 10$, and is only 20 if we take $l = 10^{-3}$. It is thus, as with all logarithmic singularities, a very weak one. It is tempting to add that this also accounts for the fact that experimenters are able to determine apparent contact angles. This issue will be taken up in the next section.

A further question arises when it is assumed that molecules from the interfacial region between the two fluids, or from the bulk of either fluid, are brought into the surface fluid layer adjacent to the solid surface, or *vice versa.* In Shikhmurzaev's theory, a finite relaxation time (larger than 1 μs) is involved before thermodynamic equilibrium is achieved. A crude estimate of this relaxation time can be obtained by supposing that it is determined by the time taken for a molecule to diffuse across the anomalous layer. Using the order-of-magnitude relation

$$t = h^2/D \tag{4}$$

for diffusion time, where h is the thickness of the surface or interfacial layer and D is the relevant diffusivity, we get $t = 10^{-5}$ s for $h = 10^{-7}$ m and $D = 10^{-9}$ m^2/s. At $U = 10$ m/s this gives a distance d from the contact line as large as 10^{-4} m over which equilibrium is attained. Even if $h = 10^{-8}$ m, $d = 10^{-6}$ m, which is still larger than l.

The above makes clear the case for studying inertia (rather than viscous) dominated flows, even in the region near the contact line when departures from the continuum assumptions become relevant. A detailed analysis has recently been given by Cox [3] which does this in terms of the Navier boundary condition 1 and other slip conditions, thus retaining a pristine continuum approach. This comes to apparently rather different conclusions from those reached by Pearson [8] as far as flow on the receding side of the interface near the contact line is concerned. This matter is discussed in more detail in the following section.

2 Contact angle and outer flow fields

In hydrostatics, the equilibrium position of an interface between two fluids lying on a locally flat surface is determined, near a contact line, by density, interfacial tension and a contact angle; in Gibbs/Young theory the value of this equilibrium contact angle is uniquely related to the interfacial tension between the two fluids and the surface energies of the two fluid/solid interfaces. We note in passing that these surface energies are suggestive of (consistent with) the notion of an anomalous layer across which slip can apparently take place.

We note at once that this notion of a contact angle is meaningful only in a continuum context. The value that it takes, which is the angle between the solid surface and the fluid interface at their intersection, the contact line, is approached, is itself in principle dependent on the averaging processes implied by the continuum approximation considered. In the equilibrium hydrostatic case ($Ca = 0$) this is not a significant issue, because as Table 1 shows the capillary length scale at which gravity forces balance interfacial tension forces is many orders of magnitude larger than the wavelength of light, the latter being the relevant length scale for ultimate visual (microscopic) estimation of the contact angle; this means that the interface is effectively flat at observable lengths.

However, as the capillary number rises towards unity (actually for any non-zero value of Ca), the radius of curvature of the interface can, in order of magnitude terms, decrease as r^{-1} and so a limiting contact angle need never exist. It is for this reason that most analyses refer to the case $Ca \ll 1$, and in formal asymptotic theory to the case $Ca \to 0$. Ultimately, of course, the molecular nature of the fluids concerned robs the concept (of radius of curvature) of any fundamental significance. This simple fact has its analogue in the uncertainties associated with precursor or pre-wetting films that are at most a few molecules thick.

It should not be expected, therefore, that phenomena characterising the chemical physics of wetting and de-wetting be adequately expressible in the simplistic large-scale continuum models of fluid mechanics, and hence elaborate constitutive behaviour of non-equilibrium interfacial layers such as those introduced by Shikhmurzaev may well be needed to maintain a continuum approach.

We now turn to the question of the influence of the outer flow field (for values of r much greater than any of those introduced by molecular considerations or surface asperities, and for which the relevant Reynolds number is much greater than unity) on the inner flow field continuum representation of these effects. Specifically, we have to consider the significance of an apparent contact angle θ_a and of its functional relation to the many parameters, both constitutive (descriptive of the materials involved) and imposed (descriptive of the outer flow field), defining the process.

Most earlier work ([2],[3],[4] and [5]) has implicitly assumed that at a small enough scale (i.e. sufficiently close to the contact line) the equilibrium contact angle θ_c correctly describes the underlying molecular physics. Dependence of an experimentally observed (apparent) contact angle (Ngan& Dussan V. [7]) upon contact line speed U (conveniently non-dimensionalised as capillary number

Ca, see equation 2 above) has been modelled in terms of some asymptotically valid matching process linking an outer flow region (characterised by a constant surface tension interface boundary condition, a no-slip solid surface boundary condition and a constant fluid viscosity) to an inner region that takes account of departures from the traditional continuum formulation. The limiting process concerned involves $Ca \to 0$, with the Ca^1 term in the outer flow asymptotic expansion providing the necessary value for $\theta_a\,(Ca)$. Definition of inner and outer regions in practice requires the specification of length scales, R for the outer flow and s say (or d) for the inner flow. This gives a dimensionless group

$$\varepsilon = s/R \tag{5}$$

which formally can become arbitrarily small. The direct matching process leads to a relation

$$\theta_a - \theta_c = O(Ca \ln \varepsilon^{-1}) \tag{6}$$

which causes no problems if ε is held fixed as $Ca \to 0$, but leads to the need for an intermediate region if $Ca \ln \varepsilon^{-1}$ is held O(1) as $Ca \to 0$.

There are clearly many ways of interpreting this curious result, but if we take a real set of three materials, we are led most naturally to consider $\mu, \rho, \sigma, \theta_c$ and s as fixed, with R and U (the imposed or operating conditions) as variable. In practice we are usually led to fix R and vary U, as for example by pulling a stretched belt into or out of a fluid interface in a given apparatus at various speeds. This varies Ca without altering ε. We note in passing that we cannot in such a system vary Ca and Re independently, as their ratio will be fixed. This is why Cox [3] uses the parameter

$$\Gamma = Re/Ca = R\sigma/\mu\nu \tag{7}$$

together with ε and θ_c to define his system. If Γ is large enough, then Re can be large while Ca is small.

One of the most interesting results of the small Re case is that the full details of the outer flow are not needed in order to get an asymptotic shape for the free surface as the contact line is approached. The question to be addressed here is whether the same holds true as Re becomes large, i.e. whether there is, for given materials and steady motion of the contact line, some universal flow field in an intermediate high-Re region which matches both the inner region and the full outer flow field, and allows an apparent contact angle still to depend upon Ca, Γ, ε and θ_c alone. This matter was raised independently by the late Raymond Cox, but not fully resolved in an unpublished manuscript left by him, see [3].

3 Lowest order outer flow solutions - model problems

The leading term ($Ca \to 0$) in earlier work is based on having a plane interface between the two mobile phases which intersects the solid surfaces at an angle

θ_c (or $\pi - \theta_c$ as seen from the other fluid). Deformations to this planar shape arise at higher, O(Ca), order. We therefore consider as our base problem that illustrated in figure 1.

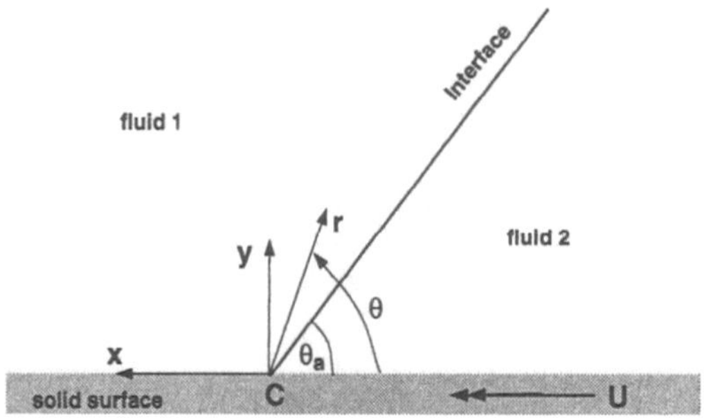

Figure 1: The steady moving contact line as seen in coordinates fixed in the contact line.

As drawn there is no length scale shown. Conventionally, an outer length scale R is used to make lengths dimensionless, while U does the same for velocity. In our case we shall use the length ν/U $(= \mu/\rho U)$ as the non-dimensionalising length; this has the added advantage of making the steady Navier-Stokes equations

$$\mathbf{u}.\nabla\mathbf{u} = -\nabla p + \nabla^2\mathbf{u} \tag{8}$$

parameter free, where the unit of pressure is ρU^2. Obviously a unique choice for μ and ρ can only be made if one of the fluids is an insubstantial gas. Otherwise, two ratios, $\rho_1/\rho_2$and μ_1/μ_2 are added to the problem specification.

We envisage, at this stage, that the entire driving force for the flow in fluids 1 and 2 (at least one of which must be a liquid) be provided by the motion of the plate $y = 0$. Boundary conditions have to be applied at the interface $\theta = \theta_a$ between fluids 1 and 2. If flow on the two sides is to be realistically coupled, then the velocity and the shear stress have to be continuous across the interface, i.e

$$\left.\begin{aligned} \mathbf{u}_1 = \mathbf{u}_2 = f(r)\widehat{\mathbf{r}} \\ \\ \mu_1\partial(\mathbf{u}_1.\widehat{\mathbf{r}})/\partial\theta = \mu_2\partial(\mathbf{u}_2.\widehat{\mathbf{r}})/\partial\theta \end{aligned}\right\} \tag{9}$$

with $\widehat{\mathbf{r}}$ the unit radial vector at the interface.

We follow Cox [3] in making first fluid 2 (the advancing contact line problem) and then fluid 1 (the receding contact line problem) an insubstantial gas.

3.1 Advancing contact line

It is fairly obvious that at high Reynolds number (Sakiadis [10]) a boundary layer solution exists for large r which has zero thickness at $x = 0$ and grows, in thickness and mass flow, as $x^{1/2}$ and so there is an outer inviscid and irrotational flow field of strength $r^{-1/2}$, which will tend to a constant pressure at large distances r. Equally, as $r \to 0$, we see that the flow will be dominated by this quasi-universal singular boundary layer flow.[1] Sufficiently close to $r = 0$, this solution must match a low-Reynolds-number flow. In [3] Cox has shown how the use of surface boundary conditions that remove the force singularity at the contact line (e.g. in equation 1 above) can be introduced into the Sakiadis analysis altering the detail of the boundary-layer flow for r of order unity, which will match without singularity onto a low-Reynolds-number flow of Huh-Scriven type.

3.2 Receding contact line

This situation is clearly not so straightforward because any fluid dragged into the contact line corner by the moving plate will tend to accumulate there. Away from the plate the flow will be everywhere almost inviscid but it will not be irrotational. Cox [3] only refers to this problem very briefly, arguing (in his §7) that the velocity field contains terms of order r and $r^{(-1+2\pi/\theta_c)}$ as $r \to 0$, and then appears to use the results of the boundary layer analysis in the advancing case which is relevant for large r; we confess that we do not understand his arguments and feel they may not address what happens in the outer flow correctly.

The approach that will now be taken is to look for a boundary-layer solution that applies for large r and is dominated by the no-slip boundary condition $\mathbf{u} = U$ on the flat plate along the line $y = 0, x < 0$. The boundary conditions at the interface $\theta = \theta_a$ must also be specified. From equation 9 we get

$$u_\theta = 0, \;\; \partial u_r / \partial \theta = 0 \tag{10}$$

Two other model problems have also been considered:

(i) that of a scraper blade moving over the solid surface; in coordinates fixed in the scraper blade this gives

$$\mathbf{u} = \mathbf{0} \quad on \quad \theta = \theta_a \tag{11}$$

(ii) that of a moving band that turns sharply at $r = 0$ and moves away from the corner along the line $\theta = \theta_a$; this gives

$$u_r = 1, u_\theta = 0 \quad on \quad \theta = \theta_a \tag{12}$$

The advantage of the latter two problems is that they are fully specified; Ca does not arise, though the "interface" is still plane. If the solution for all these cases, (10)-(12), is qualitatively similar, then we can argue that the introduction

[1]The pressure scale for the flow field above is given by ρU^2, which is not necessarily as large as that characterising an arbitrary outer flow field. In the latter case, where there is a vigorous outer flow, we would have to seek a complete solution for the flow, in which the motion of the contact line would only play a minor part.

of a second fluid, fluid 1, will not alter the qualitative similarity. In all cases we shall seek solutions in which $p \to const$ as $r \to \infty$, without justification or comment at this stage.

4 Model analysis

In terms of the planar stream function ψ, the full Navier-Stokes equations can be combined to give

$$\nabla^4\psi + \frac{1}{r}\frac{\partial(\psi, \nabla^2\psi)}{\partial(r,\theta)} = 0 \tag{13}$$

with boundary conditions

$$\theta = 0, \qquad u_r = \frac{1}{r}\frac{\partial\psi}{\partial\theta} = -1, \quad u_\theta = -\frac{\partial\psi}{\partial r} \tag{14}$$

$$\theta = \theta_a, \qquad \textit{either (i)} \quad u_\theta = \partial u_r/\partial\theta = 0 \tag{15}$$

$$\textit{or} \quad \textit{(ii)} \quad u_\theta = u_r = 0 \tag{16}$$

$$\textit{or} \quad \textit{(iii)} \quad u_\theta = 0, \quad u_r = 1 \tag{17}$$

We note that the boundary condition (14) and any one of (15)-(17) are apparently appropriately fitted by a stream function of the form

$$\psi = rF(\theta) \tag{18}$$

On substitution into (13) we get

$$(F^2 + FF'')' + \frac{1}{r}(F + 2F'' + F^{iv}) = 0 \tag{19}$$

which shows that the representation (18) can only be relevant asymptotically for very large or very small r. The latter case is the Stokes limit of Huh & Scriven [6].

4.1 Euler region

It is the significance of the outer solution given by

$$(F^2 + FF'')' = 0 \tag{20}$$

that we wish now to examine. This has a first integral

$$(F^2 + FF'') = -C = r\frac{\partial p}{\partial r} \tag{21}$$

Simple physical considerations suggest that $C > 0$, and substitution in the full Euler equation shows that

$$p = -C\ln r \tag{22}$$

If we use a scaling factor and write

$$G = C^{1/2} F \tag{23}$$

then a universal equation

$$G(G + G'') = -1 \tag{24}$$

is obtained devoid of constants. However the boundary conditions (14) become

$$G(0) = 0, \quad G'(0) = -C^{1/2} \tag{25}$$

and (15), (16) and (17)

$$\left.\begin{array}{l} (i)\; G(\theta_a) = 0,\; G''(\theta_a) = 0 \\ (ii)\; G(\theta_a) = 0,\; G'(\theta_a) = 0 \\ (iii)\; G(\theta_a) = 0,\; G'(\theta_a) = C^{1/2} \end{array}\right\} \tag{26}$$

It can be seen that the integration constant has now been transferred into the boundary conditions which, because the Euler equation is of lower order than the Navier-Stokes equation, now overdetermine the solution. It is usual to relax the condition on G' (or G'') for the Euler approximation and to introduce boundary layers near $\theta = 0$ and $\theta = \theta_a$ to satisfy the no-slip or no-stress conditions.

Equation 24 can be further integrated to give

$$G' = (\ln G^2 - G^2 + K)^{1/2} \tag{27}$$

with K a constant. We note at once that $dG/d\theta$ is singular whenever $G \to 0$. But conditions (25) and (26) require that $G = 0$ at the planar boundaries, so we are led to expect that the inflow or outflow from the boundary layers will be sufficient to remove the singularities.

Mass conservation tells us that $G' = 0$ for some value θ_0, $0 < \theta_0 < \theta_a$ and (27) that G' will be antisymmetric about θ_0, and that $|G|$ will be bounded above by $-G_0 = -G(0) > 0$. So we can integrate (27) to give

$$|\theta - \theta_0| = \int_{G/G_0}^{1} \frac{dz}{(1 - z^2 + 2\ln z/G_0^2)^{1/2}} \tag{28}$$

where G_0 and θ_0 are the two constants of integration.

If we try to impose the boundary conditions on $G(0)$, $G(\theta_a)$ directly we find that

$$\theta_0 = \frac{1}{2}\theta_a = \int_{0}^{1} \frac{dz}{(1 - z^2 + 2\ln z/G_0^2)^{1/2}} \tag{29}$$

with G' singular at the boundaries. We note that the advancing problem was solved at this order by $G \equiv 0$, $C = 0$.

We shall not pursue the relation between θ_a and G_0 any further at this point because any possible matching to a boundary layer solution will destroy, at some order, the symmetry relation (29) for cases (i) and (ii) in (26). The fully antisymmetric case (iii) will also cause difficulty unless boundary layers leaving and entering the corner region can possess the required symmetry (which subsequently we show they cannot). It is to the possibility of boundary layers, as indicated schematically in figure 2, and implied by Cox's work [3] that we now turn.

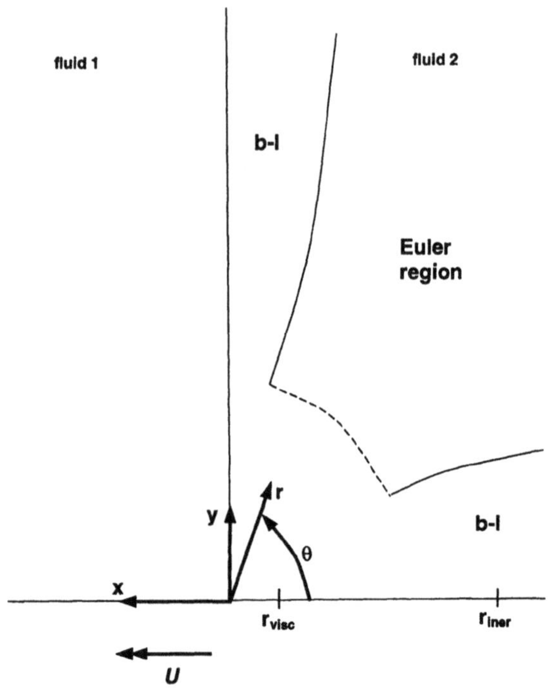

Figure 2: Diagram of coordinates and regions defined by asymptotic analysis for steady flow

4.2 Boundary layer regions

Following standard boundary layer theory, we now write

$$\psi = (-2x)^{1/2} f(\zeta),\ \zeta = y/(-2x)^{1/2} \tag{30}$$

where

$$\left.\begin{aligned} u_x &= -\frac{\partial \psi}{\partial y} = -f'(\zeta) \\ u_y &= \frac{\partial \psi}{\partial x} = -\frac{1}{(-2x)^{1/2}}\{f(\zeta) - \zeta f'(\zeta)\} \end{aligned}\right\} \tag{31}$$

for flow on the plate $y = 0$ in liquid 2.

Substitution in the boundary layer equation

$$u_x \frac{\partial u_x}{\partial x} + u_y \frac{\partial u_x}{\partial y} = -\frac{\partial p}{\partial x} + \frac{\partial^2 u_x}{\partial y^2} \tag{32}$$

yields

$$f''' + ff'' + 2C = 0 \tag{33}$$

where we have used

$$-\frac{\partial p}{\partial x} = \frac{\partial p}{\partial r} = -\frac{2C}{(-2x)}$$

to match the outer solution for pressure, with boundary conditions

$$f(0) = 0, \; f'(0) = 1 \tag{34}$$

at the moving wall (no slip).

The solution to (33) has to match onto the Euler solution, and so $f''(0)$ has to be chosen so that the term $f'''(\zeta) \to 0$ as $\zeta \to \infty$, i.e. for $\theta \gg r^{-1/2}$. We note that C in (33) could equally well have been incorporated into $f = C^{1/2}g(C^{1/4}\zeta)$ with $g'(0) = C^{1/2}$ in (34).

The more usual boundary layer equations applied to the boundary $x = 0$ with $(-x, y)$ in (30) going to $(y, -x)$ gives a different equation

$$f''' - ff'' + 2C = 0 \tag{35}$$

with the same boundary condition (34) for problem (iii) in (26). This is precisely the boundary layer equation that is relevant (with $C = 0$) in the advancing problem. It is at once obvious that simple asymmetry as implied in the discussion of the Euler solution cannot arise in the entering (decelerated) and leaving (accelerated) boundary layers.

So far, no matching between solutions of (33) and (21) has been achieved, nor does it appear from other work (S.J.Cowley, private communication) that it need exist. Instead, it was decided to turn to numerical Navier-Stokes solvers to decide the issue.

5 Numerical solution of model problems

At the end of section 2 we raised the question of whether a universal intermediate high Reynolds number solution exists for sufficiently large outer-flow-field numbers *Re*. In section 4.1 we obtained an Euler solution having a particular symmetry (velocities independent of r) as a possible candidate. It was argued briefly in Pearson [8] that a full solution (henceforth the pie solution) of the Navier-Stokes equations having boundary conditions on the arc $r = 1000$ appropriate to this Euler solution is barely distinguishable from it if an appropriate value of C is chosen, except in layers lying adjacent to the plane boundaries.

We give here rather more detailed comparison between the Euler solution, the pie solution and that for a driven cavity flow (henceforth the cavity solution)

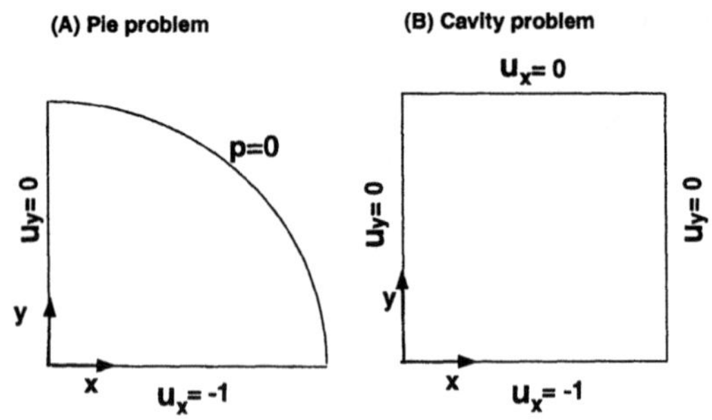

Figure 3: Schematic of domains for computational fluid dynamic calculations

in a square cavity of side length 10,000 (in dimensionles r units). The domains involved are shown in figure 3. Both pie and cavity solutions obey the planar Navier-Stokes equation 8. The relevant boundary conditions are

for the pie solution

$$\begin{aligned} u_r = -1,\ u_\theta = 0 \quad & on \quad \theta = 0,\ r \leq 1000 \\ u_r = 0,\ u_\theta = 0 \quad & on \quad \theta = \pi/2,\ r \leq 1000 \\ p = 0,\ \partial u_r/\partial r = 0 \quad & on \quad 0 < \theta < \pi/2,\ r = 1000 \end{aligned} \tag{36}$$

for the cavity solution

$$\begin{aligned} u_x = -1,\ u_y = 0 \quad & on \quad y = 0,\ 0 \leq x \leq 10,000 \\ u_x = 0,\ u_y = 0 \quad & on \quad y = 0,\ 0 \leq x \leq 10,000 \\ u_x = 0,\ u_y = 0 \quad & on \quad x = 0,\ 0 \leq y \leq 10,000 \\ u_x = 0,\ u_y = 0 \quad & on \quad x = 10,000,\ 0 \leq y \leq 10,000 \end{aligned} \tag{37}$$

Numerical solutions were obtained using AEA Technology CFX 4.1. We compare them with the Euler solution in terms of the velocity field in the region covered by the pie domain, which has its equivalent in the corner of the cavity domain according to the relation between (r, θ) and (x, y) coordinates shown in figure 1.

In figure 4 we plot the radial (u_r) and azimuthal (u_θ) velocities for $r = 100, 300, 1000$ for the pie problem; in figure 5, the equivalent velocities for the cavity problem, and in figure 6 we compare the solutions obtained with an appropriately chosen Euler solution.

In general terms the correspondence is surprisingly good. Clearly the wall boundary conditions on u_r ensure that the pie and cavity solutions for u_r depart substantially from the inviscid solution near the walls: that is why boundary-layer solutions were sought, albeit unsuccessfully, though the maxima in $|u_r|$ for both pie and cavity solutions are consistent with a $y/x^{1/2}$ scaling. The major inconsistency is in u_θ for $r = 1000$ in the cavity solution: this is not unexpected since the central vortex is restrained by three cavity faces, as opposed to just one in the pie solution; restricting attention to the region $100 \leq r \leq 300$ goes a long way towards justifying the concept of a universal intermediate asymptotic solution.

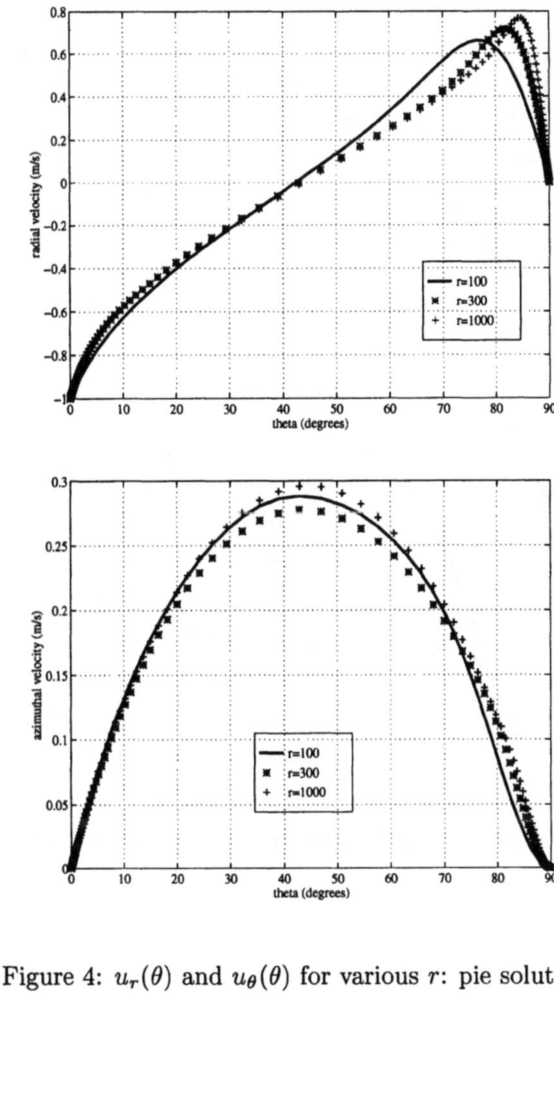

Figure 4: $u_r(\theta)$ and $u_\theta(\theta)$ for various r: pie solution

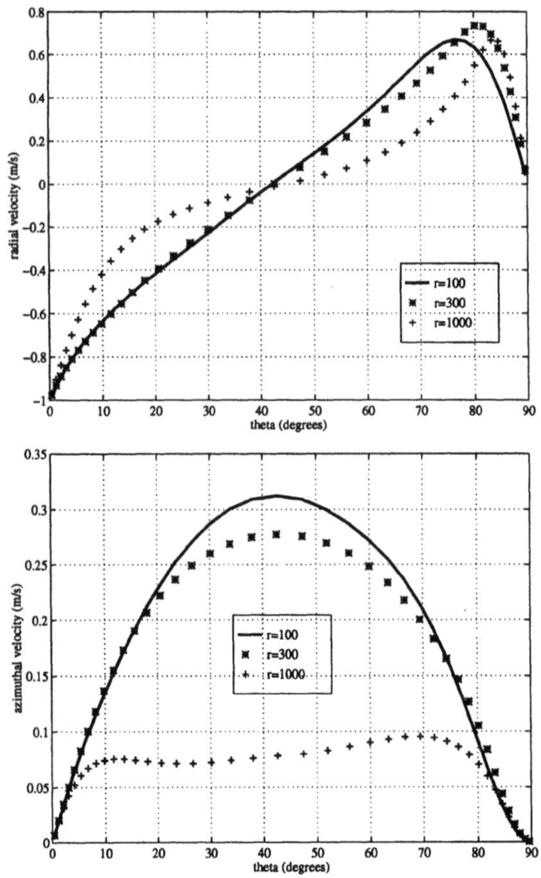

Figure 5: $u_r(\theta)$ and $u_\theta(\theta)$ for various r: cavity solution

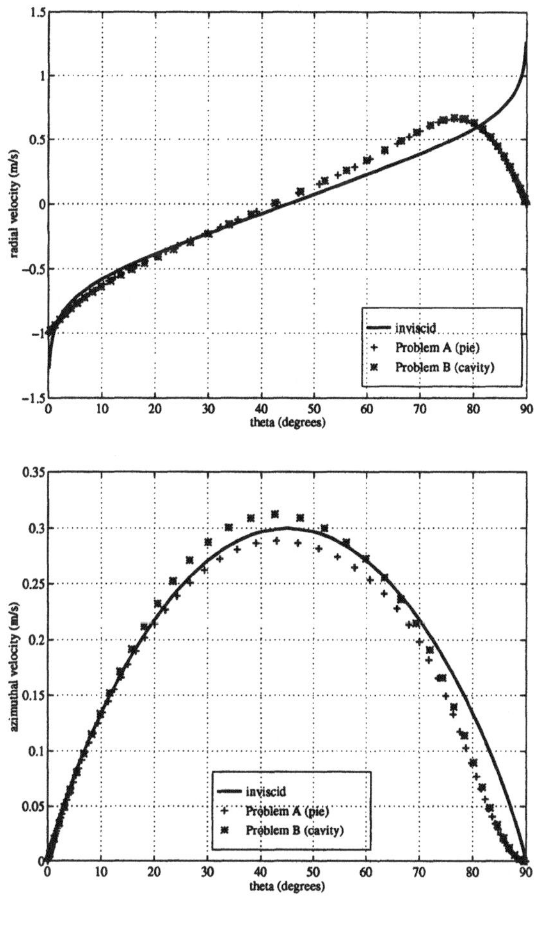

Figure 6: Comparison of inviscid and numerical solutions of Navier-Stokes equations: $u_r(\theta)$ and $u_\theta(\theta)$ for $r = 100$

6 Discussion

The Euler solution that stands at the centre of this work was introduced as a possible lowest-order approximation in r^{-1} (to the local We^0 rather than the Ca^0 term in an asymptotic expansion for the free surface problem) for rotational steady flow in an (outer) region outside the wall boundary layers. In Cox's approach, this term is put identically equal to zero. Matching to a boundary-layer solution near the walls, if achievable, would probably involve further terms in an outer expansion based on r^{-1}.

As $r \to 0$, such a joint outer/inner (boundary-layer) expansion would have to match formally to an inner Stokes expansion (over the region $r = O(1)$). It has not yet been proved that this can be done in such a way that the Stokes expansion is entirely driven by the joint outer/b-l expansion, though all earlier experience for the fixed boundary problem suggests that this would be the case. For the moving contact-line problem, however, there remains the question of how θ_a is to be obtained. The above hypothesis about a universal intermediate region (where the joint expansion would apply) independent of the outer length scale R suggests that only $\varepsilon\ Re = sU/\nu$ and Ca would be relevant to this situation.

The study was initiated by an interest in practical flows, whether associated with moving contact lines or driven cavities, and we must consider what constraints are thus imposed as outer matching conditions when $r \to \infty$ on such an intermediate solution. If the receding contact line is associated with a drop of fluid 2 immersed in fluid 1 passing over a flat plate, then we have a definite value for r at which a second, advancing, contact line passes over the same flat plate. This means that a boundary layer forms ahead of the receding contact line and so we can expect it to reach a maximum somewhere in between the two contact lines. A similar argument applies in the driven cavity flow. For both drop and cavity, we expect in a steady-state situation that there will be a net amount of vorticity (or circulation) of sign determined by the moving plate. None of this is inconsistent with the solutions developed above.

For plates of limited length in the direction of motion of a contact line we expect an unsteady flow to arise with transient motion of a contact line once it has been formed at the leading edge of the plate. This transient problem has been tackled by Billingham & King (1995) in the completely inviscid approximation; the transient nature of the problem may amke a technique for analysis of growth and decay of the boundary layers in the fluid being displaced from the plate (i.e. the receding problem) more obvious

References

[1] Billingham, J. and King, A. (1995) The interaction of a moving fluid/fluid interface with a flat plate, *J.Fluid Mech.* **59**, 707-719.

[2] Cox, R.G. (1986) The dynamics of the spreading of liquids on a solid surface, *J.Fluid Mech.* **168**, 169-194.

[3] Cox, R.G. (1997) Inertial and viscous effects on dynamic contact angles, *J.Fluid Mech.* (submitted).

[4] Dussan V., E.B. (1979) On the spreading of liquids on solid surfaces: static and dynamic contact lines, *Ann.Rev.Fluid Mech.* 371-400.

[5] Hocking, L.M. (1977) A moving fluid interface. Part 2. The removal of the force singularity by a slip flow, *J.Fluid Mech.* ***79***, 209-222.

[6] Huh, C. and Scriven, L.E. (1971) Hydrodynamic model of steady movement of a solid/liquid/fluid contact line, *J.Colloid & Interf.Sci.* **35**, 85-101.

[7] Ngan, C.G. and Dussan V., E.B. (1989) On the dynamics of liquid spreading on solid surfaces, *J.Fluid Mech.* **209**, 191-226.

[8] Pearson, J.R.A. (1995) Modelling 3-phase contact lines that move in high Reynolds number flows *Inst.Math. & Appl.* **31**, 167-171.

[9] Richardson, S. (1973) On the no-slip boundary condition, *J.Fluid Mech.* **59**, 707-719.

[10] Sakiadis, B.C. (1961) Boundary-layer behaviour on continuous solid surfaces, *AIChEJ* **7**, 221-225.

[11] Shikhmurzaev, Yu.D. (1997) Moving contact lines in liquid/liquid/solid systems, *J.Fluid Mech.* **334**, 211-250.

Mechanics

FLUID MECHANICS AND ITS APPLICATIONS

Series Editor: R. Moreau

Aims and Scope of the Series

The purpose of this series is to focus on subjects in which fluid mechanics plays a fundamental role. As well as the more traditional applications of aeronautics, hydraulics, heat and mass transfer etc., books will be published dealing with topics which are currently in a state of rapid development, such as turbulence, suspensions and multiphase fluids, super and hypersonic flows and numerical modelling techniques. It is a widely held view that it is the interdisciplinary subjects that will receive intense scientific attention, bringing them to the forefront of technological advancement. Fluids have the ability to transport matter and its properties as well as transmit force, therefore fluid mechanics is a subject that is particularly open to cross fertilisation with other sciences and disciplines of engineering. The subject of fluid mechanics will be highly relevant in domains such as chemical, metallurgical, biological and ecological engineering. This series is particularly open to such new multidisciplinary domains.

1. M. Lesieur: *Turbulence in Fluids.* 2nd rev. ed., 1990 ISBN 0-7923-0645-7
2. O. Métais and M. Lesieur (eds.): *Turbulence and Coherent Structures.* 1991 ISBN 0-7923-0646-5
3. R. Moreau: *Magnetohydrodynamics.* 1990 ISBN 0-7923-0937-5
4. E. Coustols (ed.): *Turbulence Control by Passive Means.* 1990 ISBN 0-7923-1020-9
5. A.A. Borissov (ed.): *Dynamic Structure of Detonation in Gaseous and Dispersed Media.* 1991 ISBN 0-7923-1340-2
6. K.-S. Choi (ed.): *Recent Developments in Turbulence Management.* 1991 ISBN 0-7923-1477-8
7. E.P. Evans and B. Coulbeck (eds.): *Pipeline Systems.* 1992 ISBN 0-7923-1668-1
8. B. Nau (ed.): *Fluid Sealing.* 1992 ISBN 0-7923-1669-X
9. T.K.S. Murthy (ed.): *Computational Methods in Hypersonic Aerodynamics.* 1992 ISBN 0-7923-1673-8
10. R. King (ed.): *Fluid Mechanics of Mixing.* Modelling, Operations and Experimental Techniques. 1992 ISBN 0-7923-1720-3
11. Z. Han and X. Yin: *Shock Dynamics.* 1993 ISBN 0-7923-1746-7
12. L. Svarovsky and M.T. Thew (eds.): *Hydroclones.* Analysis and Applications. 1992 ISBN 0-7923-1876-5
13. A. Lichtarowicz (ed.): *Jet Cutting Technology.* 1992 ISBN 0-7923-1979-6
14. F.T.M. Nieuwstadt (ed.): *Flow Visualization and Image Analysis.* 1993 ISBN 0-7923-1994-X
15. A.J. Saul (ed.): *Floods and Flood Management.* 1992 ISBN 0-7923-2078-6
16. D.E. Ashpis, T.B. Gatski and R. Hirsh (eds.): *Instabilities and Turbulence in Engineering Flows.* 1993 ISBN 0-7923-2161-8
17. R.S. Azad: *The Atmospheric Boundary Layer for Engineers.* 1993 ISBN 0-7923-2187-1
18. F.T.M. Nieuwstadt (ed.): *Advances in Turbulence IV.* 1993 ISBN 0-7923-2282-7
19. K.K. Prasad (ed.): *Further Developments in Turbulence Management.* 1993 ISBN 0-7923-2291-6
20. Y.A. Tatarchenko: *Shaped Crystal Growth.* 1993 ISBN 0-7923-2419-6

Kluwer Academic Publishers – Dordrecht / Boston / London

Mechanics

FLUID MECHANICS AND ITS APPLICATIONS

Series Editor: R. Moreau

21. J.P. Bonnet and M.N. Glauser (eds.): *Eddy Structure Identification in Free Turbulent Shear Flows.* 1993 ISBN 0-7923-2449-8
22. R.S. Srivastava: *Interaction of Shock Waves.* 1994 ISBN 0-7923-2920-1
23. J.R. Blake, J.M. Boulton-Stone and N.H. Thomas (eds.): *Bubble Dynamics and Interface Phenomena.* 1994 ISBN 0-7923-3008-0
24. R. Benzi (ed.): *Advances in Turbulence V.* 1995 ISBN 0-7923-3032-3
25. B.I. Rabinovich, V.G. Lebedev and A.I. Mytarev: *Vortex Processes and Solid Body Dynamics.* The Dynamic Problems of Spacecrafts and Magnetic Levitation Systems. 1994 ISBN 0-7923-3092-7
26. P.R. Voke, L. Kleiser and J.-P. Chollet (eds.): *Direct and Large-Eddy Simulation I.* Selected papers from the First ERCOFTAC Workshop on Direct and Large-Eddy Simulation. 1994 ISBN 0-7923-3106-0
27. J.A. Sparenberg: *Hydrodynamic Propulsion and its Optimization.* Analytic Theory. 1995 ISBN 0-7923-3201-6
28. J.F. Dijksman and G.D.C. Kuiken (eds.): *IUTAM Symposium on Numerical Simulation of Non-Isothermal Flow of Viscoelastic Liquids.* Proceedings of an IUTAM Symposium held in Kerkrade, The Netherlands. 1995 ISBN 0-7923-3262-8
29. B.M. Boubnov and G.S. Golitsyn: *Convection in Rotating Fluids.* 1995 ISBN 0-7923-3371-3
30. S.I. Green (ed.): *Fluid Vortices.* 1995 ISBN 0-7923-3376-4
31. S. Morioka and L. van Wijngaarden (eds.): *IUTAM Symposium on Waves in Liquid/Gas and Liquid/Vapour Two-Phase Systems.* 1995 ISBN 0-7923-3424-8
32. A. Gyr and H.-W. Bewersdorff: *Drag Reduction of Turbulent Flows by Additives.* 1995 ISBN 0-7923-3485-X
33. Y.P. Golovachov: *Numerical Simulation of Viscous Shock Layer Flows.* 1995 ISBN 0-7923-3626-7
34. J. Grue, B. Gjevik and J.E. Weber (eds.): *Waves and Nonlinear Processes in Hydrodynamics.* 1996 ISBN 0-7923-4031-0
35. P.W. Duck and P. Hall (eds.): *IUTAM Symposium on Nonlinear Instability and Transition in Three-Dimensional Boundary Layers.* 1996 ISBN 0-7923-4079-5
36. S. Gavrilakis, L. Machiels and P.A. Monkewitz (eds.): *Advances in Turbulence VI.* Proceedings of the 6th European Turbulence Conference. 1996 ISBN 0-7923-4132-5
37. K. Gersten (ed.): *IUTAM Symposium on Asymptotic Methods for Turbulent Shear Flows at High Reynolds Numbers.* Proceedings of the IUTAM Symposium held in Bochum, Germany. 1996 ISBN 0-7923-4138-4
38. J. Verhás: *Thermodynamics and Rheology.* 1997 ISBN 0-7923-4251-8
39. M. Champion and B. Deshaies (eds.): *IUTAM Symposium on Combustion in Supersonic Flows.* Proceedings of the IUTAM Symposium held in Poitiers, France. 1997 ISBN 0-7923-4313-1
40. M. Lesieur: *Turbulence in Fluids.* Third Revised and Enlarged Edition. 1997 ISBN 0-7923-4415-4; Pb: 0-7923-4416-2

Kluwer Academic Publishers – Dordrecht / Boston / London

Mechanics

FLUID MECHANICS AND ITS APPLICATIONS

Series Editor: R. Moreau

41. L. Fulachier, J.L. Lumley and F. Anselmet (eds.): *IUTAM Symposium on Variable Density Low-Speed Turbulent Flows.* Proceedings of the IUTAM Symposium held in Marseille, France. 1997 ISBN 0-7923-4602-5
42. B.K. Shivamoggi: *Nonlinear Dynamics and Chaotic Phenomena.* An Introduction. 1997 ISBN 0-7923-4772-2
43. H. Ramkissoon, *IUTAM Symposium on Lubricated Transport of Viscous Materials.* Proceedings of the IUTAM Symposium held in Tobago, West Indies. 1998 ISBN 0-7923-4897-4

Kluwer Academic Publishers – Dordrecht / Boston / London

ERCOFTAC SERIES

1. A. Gyr and F.-S. Rys (eds.): *Diffusion and Transport of Pollutants in Atmospheric Mesoscale Flow Fields*. 1995 ISBN 0-7923-3260-1
2. M. Hallbäck, D.S. Henningson, A.V. Johansson and P.H. Alfredsson (eds.): *Turbulence and Transition Modelling*. Lecture Notes from the ERCOFTAC/IUTAM Summerschool held in Stockholm. 1996 ISBN 0-7923-4060-4
3. P. Wesseling (ed.): *High Performance Computing in Fluid Dynamics*. Proceedings of the Summerschool held in Delft, The Netherlands. 1996 ISBN 0-7923-4063-9
4. Th. Dracos (ed.): *Three-Dimensional Velocity and Vorticity Measuring and Image Analysis Techniques*. Lecture Notes from the Short Course held in Zürich, Switzerland. 1996 ISBN 0-7923-4256-9
5. J.-P. Chollet, P.R. Voke and L. Kleiser (eds.): *Direct and Large-Eddy Simulation II*. Proceedings of the ERCOFTAC Workshop held in Grenoble, France. 1997 ISBN 0-7923-4687-4

KLUWER ACADEMIC PUBLISHERS – DORDRECHT / BOSTON / LONDON

MIX
Papier aus verantwortungsvollen Quellen
Paper from responsible sources
FSC® C105338

If you have any concerns about our products,
you can contact us on
ProductSafety@springernature.com

In case Publisher is established outside the EU,
the EU authorized representative is:
Springer Nature Customer Service Center GmbH
Europaplatz 3, 69115 Heidelberg, Germany

Printed by Libri Plureos GmbH
in Hamburg, Germany